THE HISTORY OF THE SOVIET
ATOMIC INDUSTRY

THE HISTORY OF THE SOVIET ATOMIC INDUSTRY

Arkadii Kruglov

Translated from the Russian edition by Andrei Lokhov

Routledge
Taylor & Francis Group

LONDON AND NEW YORK

First published 2002
by Taylor & Francis

2 Park Square, Milton Park, Abingdon, Oxfordshire OX14 4RN
52 Vanderbilt Avenue, New York, NY 10017

Routledge is an imprint of the Taylor & Francis Group, an informa business

First issued in paperback 2019

Originally published in Russian in 1994 as
Как создавалась атомная промышленность в СССР by TsNIIatominform Moscow.
© A. K. Kruglov, 1994
© ЦНИИатоминформ, Moscow, 1994

British Library Cataloguing in Publication Data
A catalogue record for this book is available
from the British Library

Library of Congress Cataloging in Publication Data
A catalog record for this book has been requested

ISBN 978-0-415-26970-4 (hbk)
ISBN 978-0-367-39592-6 (pbk)

CONTENTS

Introduction

The year of 1996 witnessed the 100th anniversary of the discovery of radioactivity which gave birth to nuclear physics — scientific basis of the atomic industry. The basic discoveries made by Becquerel, Pierre and Marie Curie, Thompson, Rutherford, and their successors are reviewed comprehensively in [1].

As early as March 1896 H. Becquerel published his first paper devoted to some properties of invisible rays emitted by uranium salts. In late 1897 M. Sklodowska-Curie took interest in the newly discovered uranium or Becquerel rays. When she established the fact that uranium and thorium emitted Becquerel rays, she wrote: "I called the capability of emitting such rays the *radioactivity*[1] thus introducing a new term which has since then become conventional in science." Radiating elements were called *radioelements*, the term coined from the Latin word *radius* meaning a ray. It is commonly understood by the scientific community that Becquerel discovered radioactivity in 1896, and the year 1898 is supposed to be when M. Curie introduced the term *radioactivity*. In 1898 when studying the radiation of already known uranium and thorium compounds, the Curie couple discovered two new elements. On July 18, 1898, they reported the discovery of polonium, and on December 26, 1898, the discovery of radium. In 1899 M. Curie rejected a proposed identity between Becquerel rays and Roentgen rays discovered in 1895. She emphasized that the radioactivity phenomenon is accompanied by mass loss of the source material and permanent energy loss of the radioactive material.

M. Sklodowska-Curie was the only one among world scientists awarded the Nobel Prize twice[2]: she shared the 1903 Physics Prize with P. Curie and Becquerel for the study of radioactivity and discovery of radium, and was awarded the 1911 Chemistry Prize for the isolation of metallic radium.

Before the Russian revolution, L.S. Kolovrat-Chervinskiy, a researcher from the Saint-Petersburg University, worked at the Radium Institute in Paris. During the Soviet period, Z.V. Ershova (1937) and other Russian researchers also worked there.

D.I. Mendeleev, author of the periodic system of elements, considered the study of radioactivity of exceptional significance; he visited the Curies' laboratory in 1902. He wrote at that time: "Being sure that the study of uranium associations beginning with the natural sources will bring us to many other discoveries, I may confidently recommend those who are looking for new research objectives to consider uranium associations most comprehensively." In his opinion, this would allow the reasons for the periodicity of elements to be established in more detail; this idea was later implemented in the works of English and Russian scientists [2–4].

A theory of radioactive transmutations was proposed by E. Rutherford and F. Soddy in 1902–1903 [2]. The name of Rutherford is connected with the discovery of atomic nuclei, he is rightly considered to be a founder of nuclear physics. In 1899 he discovered the emission of alpha and beta rays, and in 1903, predicted the existence of transuranium elements. In 1908 he was awarded the Nobel Prize in Chemistry for the studies in the transmutation of elements and in the chemistry of radioactive materials. In 1911 he established the existence of positively charged atomic nuclei, approximately 10^{-12} cm in diameter. In 1919 Rutherford carried out the first artificial nuclear reaction converting nitrogen into oxygen:

$$_7N^{14} + {}_2He^4 \rightarrow {}_8O^{17} + {}_1H^1.$$

He also discovered an elementary particle, the proton, and in 1920 predicted the existence of neutrons and deuterons. Those were not the only discoveries made by the founder of nuclear physics. Rutherford was the only scientist who was a member of all academies of the world, including the USSR Academy of Sciences from 1925 [5].

Besides the discoveries of radioactive decay (1902) and radioactive transmutations (1903) made jointly with Rutherford, Soddy introduced the notion of half-life $T_{1/2}$ and formulated the law of radioactive displacement in the periodic system of elements (1911): two positions to the left in the case of α-particle disintegration and one position to the right in the case of β-particle disintegration. In 1921 Soddy was awarded the Nobel Prize.

Among the first Russian scientists who participated in research at Rutherford's laboratory in 1910–1912 were V.A. Borodovskiy (Assistant Professor, Petersburg University) and Yu.N. Antonov (researcher, Chemical Laboratory, Petersburg Academy of Sciences). In the Soviet time P.L. Kapitsa worked for Rutherford from 1921. Over a period of 13 years, Kapitsa became not only the most brilliant follower of Rutherford, but also a well-known expert in super-high magnetic fields. In 1933 Rutherford succeeded in creating a new physics laboratory at the London Royal Society founded by the Mond Foundation;

Kapitsa was appointed as its chief.[3] In 1929 Kapitsa was elected a member of the London Royal Society. Later, from 1928 to 1930, K.D. Sinelnikov worked for Rutherford. Among the trainees of the Cavendish Laboratory were A.I. Leipunskiy and Yu.B. Khariton who later became Academicians and supervisors of the leading nuclear centers of the USSR.

Shortly after the discovery of radioactivity, eminent Russian scientists N.N. Beketov, A.M. Butlerov, N.A. Morozov, N.A. Umov, and others advanced a hypotheses of the complex structure of nuclei and their possible fission [4]. In 1905, the outstanding theoretical physicist Umov wrote: "We face a new great problem: physics and chemistry of atoms − microphysics and microchemistry... Life of the atomic interior will offer us the properties and laws which might differ from those constituting the old, already ancient physics."

On the 7th November, 1907, in recognition of the great importance of the discovery of radioactivity the Petersburg Academy of Sciences elected M. Curie an associated member. In France, she was elected as a member of French Academy of Sciences 15 years later − on the 7th February, 1922.

VLADIMIR IVANOVICH VERNADSKIY,
Academician from 1912.
First Director of the Radium Institute in Petrograd (St. Petersburg). First world scientist who predicted the value of radioactive materials as a power source.

The great historian of science, chemist, and philosopher B.M. Kedrov comments on the contribution of Academician N.N. Beketov to elucidation of the nature of radioactivity and his role in the history of its discovery: "In the discovery of radioactivity it is of great importance that Becquerel was the first who empirically observed the radioactive radiation, the Curie couple experimentally discovered a new element − radium, and Rutherford, Soddy and Beketov were the first who elucidated the nature of this phenomenon which was obscure and enigmatic until that time." Academician Beketov, the oldest physicochemist of pre-revolutionary Russia, was at the front line of modern science. He wrote at that time: "Radium properties from the beginning put in doubt the existing scientific views of invariable and indivisible elements and energy conservation law, because radium permanently produces energy and material emanation... It turned out that the discovery of radium and its spontaneous release of energy and emanation would suffice to lead again to confusion between energy and matter." In his paper *Explanation of Radium Properties,* Beketov emphasized that the decay of radium into radon (emanation) and helium, accompanied by a release of considerable energy was a natural phenomenon consistent with the law of energy conservation: "Particles of radium possessing a large amount of energy (as was shown earlier) can produce low-energy substances and simultaneously release a considerable amount of energy. Thus, the generation of noble gases and the release of energy do not lead to conflicting conclusions: these two phenomena are consistent with one another" [4]. Owing to his profound understanding of radioactivity, Beketov drew a conclusion about possible artificial transmutation of atoms in specially designed facilities as far back as 1909, ten years before Rutherford carried out the first artificial transmutation of a nitrogen nucleus to oxygen. He emphasized that "the artificial transmutation of elements under external actions can proceed in two possible ways – through dissociation and synthesis," i.e., through a complete disintegration of an element or conversion of a low-energy element to a high-energy one.

L.S. Kolovrat-Chervinskiy, a student of O.D. Khvolson, worked in France for 5 years; his research during the first 3 years was supported by a Curies' grant awarded by the University of Paris. On his return to Russia, he joined the Laboratory of the Mineralogical Museum, Russian Academy of Sciences, headed, at different times, by Academicians F.N. Chernyshev, V.I. Vernadskiy, and A.E. Fersman. Kolovrat-Chervinskiy studied the temperature dependence of radon released from various salts of radium, beta-radiation, electric conductance and other properties of radon. In the early period of radioactivity research (1898–1913) nearly 30 natural radioactive isotopes were discovered by scientists who studied their mutual transmutations and radiation properties. Tables of characteristics of all radioactive materials compiled by Kolovrat-Chervinskiy were published three times during 1910–1913. These tables comprised half-life values, radiation pattern, alpha-particles path lengths, and absorption coefficients of beta and gamma-rays. The tables were widely used by scientists in many countries. In 1919, at the International Congress on Radiology and Electricity, Kolovrat-Chervinskiy presented a summary of his publications. The tables were included in the Congress Proceedings and, also, in M. Curie's monograph on radioactivity and Rutherford's monograph *Radioactive Materials and Radiation.* Later, Kolovrat-Chervinskiy worked as a head of the Radium Department of the State X-Ray and Radiological Institute created in 1918.

In pre-revolutionary Russia, the study of radioactivity and internal energy release processes was purposefully started under the guidance of Academician V.I. Vernadskiy. At the Congress of the British Association of Sciences held in 1908 in Dublin, he thoroughly studied papers published by French scientists and Rutherford, as well as publications of mineralogists unveiling a geological application of the radioactivity phenomenon. Two years later a special Radium Commission was established at the Petersburg Academy of Sciences. Among its members were the outstanding scientists: V.I. Vernadskiy, A.P. Karpinskiy,

F.N. Chernyshev, N.N. Beketov, B.B. Golitsin, and P.I. Valden. As early as 1911, the first radium surveys were carried out in Trans-Caucasus, Trans-Baikal, Fergana, and Ural regions.

Besides Vernadskiy, among the scientists who worked at the laboratory of the Mineralogical Museum were A.E. Fersman, K.A. Nenadkovich, L.S. Kolovrat-Chervinskiy, B.A. Linder, V.G. Khlopin, B.G. Karpov, and other contributors to the radioactivity, science of radium production, and uranium utilization.

In [1] Staroselskaya-Nikitina describes the visit of M. Curie to Krasnoyarsk and East Siberia region located close to the *Krugobaikal* railroad in 1914, just before the World War I. The Siberian local historians and naturalists (teachers and deportees) sent samples of uranium minerals found in these areas to their compatriot, M. Curie in Paris (she was born in Warsaw, then a part of Russia). Even peasants of the Minusinsk District, Enisei Province, were among those who participated in the search for precious radium which was believed to heal cancer and sarcoma patients. I.G. Prokhorov, chairman of the Kazansk-Bogorodsk labor mining and production co-operative (Minusinsk District), who attended the lecture presented by M. Curie in Krasnoyarsk, emphasized that her visit was due to her scientific interest in the samples of uranium minerals she received in Paris.

Until 1918 radium production in Russia was concentrated in the mineralogical and radiological laboratories of the Academy of Sciences. The studies in these laboratories were personally supervised by V.I. Vernadksiy, an outstanding scientist of the 20th century. Later he recalled: "Before long, the man will have atomic energy at his disposal, a powerful source that will allow him to build his life according to his wishes. This is likely to happen within centuries. But it will take place for sure. Will the man manage to use this power for good purposes and not for self-destruction? Is he able to use the power which will be inevitably given to him by science?..." For the most part the words of the greatest Russian scientist turned out to be prophetic − the second part of the 20th century witnessed the creation of nuclear weapons, the atomic and nuclear power industry.

Radioactive sources have long been used in many vital fields of man's activity, particularly in medicine. Only half a century passed since the mastering of atomic power, and some 17% of world power is currently produced at atomic power plants (APP). However, man is still learning how to manage this power and to provide safe and environmentally acceptable handling of nuclear fuel and radioactive waste. Only countries with a highly developed industrial and science base were able to participate in atomic energy projects, the former USSR included. Not only were sophisticated nuclear weapons created, but also the atomic industry and reactor building industry for nuclear power production were developed. Russia is experienced in the building and operation of the complete system of facilities involved in the nuclear fuel cycle. Its atomic industry potential can be compared only with that of the USA. The secrecy imposed on all phases of the USSR atomic project resulted in the present-day lack of public knowledge about the huge, mostly tragic activity aimed at mastering atomic power. That is why memoirs of those who participated in the nuclear weapons project and the pioneers of the USSR atomic industry are of special interest today. Their experience allows us to recognize both the danger and tremendous capabilities of atomic power and its role in our life more objectively and without dramatic effects.

In 1992 the Ministry of Atomic Industry (Minatom) of Russia decided to prepare an open publication of the comprehensive history of the atomic industry in the former USSR and Russian Federation − of the history which until recently was one of the most secret fields. This atmosphere of secrecy was responsible for numerous publications by incompetent authors and the circulation of tendentious information about problems related to the utilization of nuclear power. Professor O.N. Rusak, President of the Association of experts in human safety (St. Petersburg), analyzed more than one thousand papers on the Chernobyl accident published in the Ukraine in the second half of 1991. He found out that only 30 of them were accurate. A similar situation is observed in the local press in many regions of Russia. Within 50 years of beginning activities aimed at developing civil and military projects in the domestic atomic industry many designers and workers were involved in some tragic accidents and even nuclear catastrophes. However, because of the unnecessary secrecy, scientists, specialists and the public do not have objective information, especially about the hazardous effects of various atomic production facilities on the environment and human health. It should be noted that for the same reason public have little knowledge about the role of the atomic industry in establishing the standards for radiation protection of personnel, natural resources, and the population. Also, little is known about the contribution of particular specialists and teams of research workers from various institutes and organizations to the solution of atomic problems. In recent years new publications have appeared devoted to specific problems of the atomic project. The publication of the history of the atomic industry will allow one to assess the expediency of particular research projects and recognize the significance of timely initiation of new trends in science, and will clarify the contribution of various atomic industry enterprises to the national economy. The development of atomic industry was accompanied by the formation of construction and production associations, foundation of research and design institutes, and new branches of the national industry. The 1990s were marked with a series of jubilees in the history of the national atomic industry:

-    70 years since the foundation of the Radium Institute by V.I. Vernadskiy;
-    60 years since the advent of the modern nuclear physics (dated since the discovery of neutron by J. Chadwick, Great Britain, on February 17, 1932);

- 60 years since a hypothesis of the proton-neutron, electron-free composition of atomic nucleus was proposed by Soviet physicist D.I. Ivanenko independently of W.K. Heisenberg (Germany), and since protons and neutrons were recognized as elementary particles (August 1932);
- 50 years since the USSR State Committee of Defense (GKO) made a decision to renew research aimed at the study of atomic energy released from the fission of uranium nuclei (late 1942), the work that was interrupted by the war;
- 50 years since I.V. Kurchatov was appointed a head of atomic energy utilization projects and since the organization of Laboratory No. 2, Academy of Sciences of the USSR (March 10, 1943);
- 50-years activity of the leading institutes of atomic industry − NII-9 (All-Union Research Institute of Inorganic Materials − VNIINM), Thermal Engineering Laboratory (TTL), later called the Institute for Theoretical and Experimental Physics − ITEF), Efremov Research Institute for Electrophysical Tools (NIIEFA), and other institutes and design bureaus located in Moscow, Leningrad, Ural region, Gorky, Podolsk, Obninsk, and Arzamas.
- August 20 and 30, 1995 − 50 years since the Special Committee and the First Chief Directorate (PGU) under the USSR Peoples Commissariat (Sovnarkom) were established to solve the problems involved in the development of nuclear weapons and the creation of an atomic industry. These organizations were given authority to draw in their work institutes, design bureaus, and production facilities from any branches of the industry; they provided coordination and control of all research and engineering projects dealing with the atomic problem. Efficient management and mobilization of all necessary resources inside the country made it possible to construct the first research reactor F-1 at Laboratory No. 2 and initiate a controlled chain nuclear reaction as early as December, 1946. By July, 1948 the first industrial reactor intended for production of weapon plutonium was designed, constructed, and put to production at Chelyabinsk-40 (named successively as Base No. 10, Center No. 817, Mendeleev Center, and a Mayak production association). The first radiochemical plant intended for extraction of plutonium from irradiated uranium and the plant producing the plutonium product were put into operation within half a year. As early as December 22, 1948, the first batch of irradiated materials intended for the extraction of plutonium arrived at the radiochemical plant. The weapon components ordered by the affiliate of Laboratory No. 2 (Design Bureau KB-11[4] were fabricated from the refined plutonium, and the first USSR atomic plutonium bomb was manufactured, which was successfully tested at the Semipalatinsk Site on August 29, 1949. This bomb was the first in the USSR, but not the first in the world. The Hiroshima and Nagasaki events took place

before this test. After that time it became clear that the fate of these cities would not be repeated elsewhere without an adequate response.

The active production of plutonium was carried out under extremely unfavorable conditions. Because of the lack of information about the hazardous effects of radiation on man's health and natural environment, imperfect technology and equipment, many workers and scientists have been exposed to high levels of radiation. Most of them were assigned to the category of professional staff and were temporarily suspended from the basic production jobs. Some of them were transferred to safe industrial branches, thus having lost appropriate medical observation. Moreover, vast areas adjacent to a number of nuclear centers were contaminated with radionuclides.

For a number of reasons to be considered below the artificial element − plutonium, rather than ^{235}U abundant in nature, was used in the first nuclear explosive device. The theory of building an explosive device using ^{235}U was substantiated as early as 1938–1939. Plutonium was discovered only in 1940, and its isotope ^{239}Pu used in weapons for the most part was obtained later.

In the past, leaders of the atomic industry tried to make the results achieved by particular teams and specialists public. In 1967 M.G. Pervukhin proposed to lift the secrecy veil from the atomic project. In early 1943, according to a decree issued by the GKO State Committee he was assigned to arrange the first teams to conduct atomic energy research for military purposes. Together with I.V. Kurchatov he was a founder of Laboratory No. 2, and coordinated works conducted by other research and production teams engaged in solving the atomic problem. On May 31, 1967, he handed to the Politburo of the KPSS Central Committee a note, in which he suggested the publication of a collection of papers and memoirs by atomic industry veterans. To justify his proposal he wrote: " ...the atomic industry, and atomic and thermonuclear bombs have been created during a fairly short time. Unfortunately, many of those who started and organized this giant project have gone (I.V. Kurchatov, V.G. Khlopin, B.L. Vannikov, A.P. Zavenyagin, V.A. Malyshev, P.M. Zernov, to name but a few)." Nonetheless, this proposal was neglected [6–8].

In 1984 the USSR Union of Writers proposed to write a monograph about the history of atomic industry in the USSR. Under the decision of Minister E.P. Slavskiy, writer A.A. Logvinenko was permitted to visit the leading research centers and interview veterans of atomic industry and eminent scientists. Unfortunately, this book was never written. Now less than half of the forty scientists and specialists whose memoirs were supposed to have been included in that book are still alive.

The books *Arzamas-16* by V.S. Gubarev, *First Atomic Plant: Notes of a Research Engineer* by V.I. Zhuchikhin, *Arzamas-16 Nuclear Center: Pages of History* by S.G. Kochryants and N.N. Gorin (published by the Atomizdat, Moscow and VNIIEF, Arzamas) deal only with the initial period of research conducted in KB-11 (Affiliate of Laboratory No. 2) during creation of nuclear weapons. In these publications much attention is given to the memoirs of scientists and designers of the Russian Federation Nuclear Center (VNIIEF) and their business contacts with the directors of the uranium project. Zhuchikhin described in detail site preparation and testing of the first atomic bomb at the Semipalatinsk Site.

In 1933 the Editorial Board of the Bulletin of the Public Information Center published the first, journal version of the present book under the title *On the History of Atomic Science and Industry,* which described by whom and under what conditions the first atomic industry production centers, institutes, and design bureaus were established.

The Plutonium Center No. 817 in South Ural was an outpost of our atomic industry. This Center included uranium-graphite and heavy-water nuclear reactors, a radiochemical plant for the extraction of plutonium extracted from decay products, and a plant for manufacturing basic fissile material components for an atomic bomb. In the Mid-Ural region sophisticated production centers were built for obtaining another fissile component for an atomic bomb — 90%-enriched ^{235}U. The first factory for uranium mining and production was put into operation in Central Asia to supply the atomic centers of South and Central Ural with natural uranium. Production of uranium hexafluoride was arranged to extract ^{235}U from natural uranium. To produce plutonium facilities were installed to fabricate pure metallic uranium billets to be used in nuclear reactors. The first (in Europe and Asia) controllable nuclear chain reaction could not be carried out under the guidance of Kurchatov at Laboratory No. 2, and the first industrial uranium-graphite and heavy-water nuclear reactors could not be constructed at Laboratory No. 3 (later successively named TTL and ITEF), headed by Alikhanov, without arranging sophisticated production facilities for special graphite, heavy water, and other materials at plants belonging to other industries.

This book describes the efforts of scientists, designers and builders who ensured the construction of the first atomic industry production centers in 1946–1949. Due to lack of space it has been impossible to mention many contributors from the Academy and research institutes and design bureaus of the Mintsvetmet, Minkhimprom, and Mintyazhmash Ministries and other industries who participated in the development of technological processes, new materials, and special-purpose equipment. Considerable attention is given to the regular service staff and some technological service structures of the first atomic centers. The decisive role of the Special Committee and the PGU in the uranium project is emphasized.

The description of the work done under the atomic project before, during and after the war (late 1940s) is important for understanding the reasons that forced the USSR to join a more than 50-year nuclear armaments race. The comparative analysis of atomic research trends in the USA, in war-time

Germany, and in the USSR gives an objective insight into the USSR atomic industry which was basically created for defense purposes.

I hope that the reader will understand that many of those who designed and serviced sophisticated nuclear facilities are not mentioned in this book for lack of information or other reasons.

NOTES

1   Some authors ascribe the introduction of the term *radioactivity* to H. Becquerel, others to E. Rutherford.
2   Edit. Rem.: Linus Carl Pauling was another person who won two Nobel Prizes (1954 and 1962).
3   From 1934 P.L. Kapitsa worked in the USSR. In 1935, after approval by E. Rutherford, the USSR bought unique equipment from the Mond Laboratory for £30,000 to study super-high magnetic fields and super-low temperatures. From 1935 to 1946 Kapitsa worked as the director of the Institute of Physics Problems (IFP) he founded in 1935.
4   It was successively named the Volga Office, KB-11, Object No. 550, "Kremlev," "Center 300, Moscow," Arzamas-75, and Arzamas-16 (Sarov), the All-Russian Research Institute of Experimental Physics, VNIIEF.

CHAPTER 1

Prewar Period (1918–1941)

Preliminary research laying the foundation of the atomic industry was conducted well before World War II. At that time studies of physical and chemical effects of radioactivity were in progress at university laboratories and newly created institutes of the USSR Academy of Sciences in Leningrad, Moscow, Kharkov, Odessa, Tomsk, and other cities. In 1922 V.I. Vernadskiy emphasized that "radioactive phenomena present to us the source of atomic energy which is a million times more powerful than any other sources one could imagine." Representatives of the USSR research centers worked as probationers at the laboratories headed by M. Curie, E. Rutherford, N. Bohr, and other great scientists of Western Europe. All great scientific discoveries and advances made at that time at home and abroad were covered in detail in monographs and popular scientific publications [2–4].

It is interesting to recall fundamental researches and their basic phases dealing with the atomic problem both in Russia and highly developed countries (Germany, Great Britain, France, and USA). Of particular importance is the state of development of the German Uranium Project achieved by the middle of 1941, because it was the primary reason for the subsequent atomic boom, first in Great Britain, and then in the USA. The advances in nuclear physics in the USSR are described comprehensively in A.P. Aleksandrov's paper published in the jubilee collection [9]. The author describes the problems with the creation of atomic engineering devices at a later period (after 1940) within the limits of information available at that time. Scientific advances in nuclear physics and other fields achieved at the Radium Institute of the Academy of Sciences (RIAN), Leningrad and Ukraine Physical Engineering Institutes (LFEI and UFEI) were considered in detail in jubilee monographs published at that time [3, 10–12].

The year of 1932 is known in the history of science as the *radioactivity annus mirabilis* [2]. It was marked by the discoveries of neutron, positron, and deuterium made at the laboratory headed by E. Rutherford. On May 2, 1932, Cockroft and Walton from Rutherford's laboratory were the first to split the lithium, boron, and aluminum nuclei using accelerated protons. Later the same year, Soviet physicists K.D. Sinelnikov, A.I. Leipunskiy, A.K. Walter, and G.D. Latyshev from UFEI (Kharkov)[1] also split the lithium nucleus [10]. This was only a repetition of the Cockroft and Walton experiment. However to accomplish this, they solved a complicated engineering problem: a high voltage source was constructed consisting of a specially-designed vacuum tube where the beam of accelerated particles was generated.

In the telegram addressed to Stalin, Molotov, and Ordzhonikidze, which was published in the newspaper *Pravda* on October 22, 1932, I.V. Obreimov, Director of the UFEI Institute wrote: "On the 10th of October, the UFEI researchers were the first in the USSR and second in the world who succeeded in splitting the lithium nucleus by bombarding it with hydrogen nuclei accelerated in a vacuum tube. Advances of the Institute open tremendous opportunities for studying the structure of atomic nuclei... The UFEI Institute constructs a more powerful facility for splitting other elements." Thus, it was obvious that to provide an effective study of nuclear structure one needs the sources of neutral particles − neutrons, and various radiation sources, both natural (radium, radon) and artificial, such as beams of high-energy charged particles generated in specially-designed accelerators.

In 1932, at the session of the RIAN Scientific Board chaired by V.I. Vernadskiy, a decision was made to build a cyclotron at RIAN. At the LFEI Institute a research group was organized to study the atomic nucleus.[2] In the same year, G.A. Gamow, a worker of the RIAN Institute[3] published a monograph *Structure of the Atomic Nucleus and Radioactivity*. In fact it was the first attempt to interpret the nuclear processes [3].

In 1935 Kurchatov published his monograph *Splitting of the Atomic Nucleus* [13] where he analyzed the results of foreign experiments in alpha-splitting of the copper to uranium nuclei based on Gamow's theory of the nucleus field structure after scattering of alpha-particles. In this monograph much attention was given to the analysis of theoretical and experimental papers dealing with the interaction of neutron with nuclei of various elements. The isotope uranium-235, the basic constituent of nuclear fuel, was discovered in natural uranium by F. Dempster (USA) in the same year.

In 1935 I.V. Kurchatov, B.V. Kurchatov, L.V. Mysovskiy, and L.I. Rusinov from the LFEI and RIAN Institutes discovered the nuclear isomerism of radioactive isotope ^{80}Br [14]. This discovery was preceded by the discovery of isotopes and isobars [2].

In 1938–1939 the German researchers O. Hann and F. Strassmann found that when the uranium nucleus absorbed a neutron it split and produced a large amount of energy. A number of scientists, E. Fermi and F. Joliot-Curie being first among them, came to the conclusion that the uranium nucleus fission produced 2 or 3 neutrons, which under certain conditions could in turn be absorbed by uranium nuclei. So the nuclear chain reaction results in an explosion of monstrous intensity. E. Fermi and other scientists discovered that the probability of neutrons being absorbed by ^{235}U nuclei might be hundreds of times higher, if nuclei interacted with slow neutrons instead of the fast ones. Neutrons slow down during their interactions with the nuclei of light elements (hydrogen, deuterium, carbon, etc.). In papers published by N.N. Semenov, Yu.B. Khariton, and Ya.B. Zeldovich in 1935 the possibility of a nuclear chain (explosion) reaction was established [9]. However, calculations published by Khariton and Zeldovich showed that a self-sustaining chain reaction could not take place in natural uranium if conventional water and graphite were used as neutron moderators. For these purposes uranium enriched with isotope ^{235}U was required [15]. These researchers mistakenly thought that a nuclear chain reaction could produce an explosion only given the enriched uranium and moderator: "... to provide conditions under which a chain explosion could take place, it is necessary to use heavy hydrogen or, perhaps, heavy water, or some other heavy substance featuring a sufficiently low neutron absorption cross-section as a neutron moderator."

Thus, the recommendations for selecting the quantity and composition of uranium required for a self-sustaining nuclear chain reaction were rather contradictory. It should be noted that studies carried out by the Soviet scientists at that time only improved experimental and theoretical results obtained by foreign researchers. I.N. Golovin noted later: "... the first, very approximate estimates of the size of uranium ball being heated by the fission-released energy were published by French physicist F. Perrin" [14]. According to his estimates (accurate up to cross-sections of neutron-uranium nucleus interaction available at that time), a self-sustaining nuclear reaction could take place in a 7.5-ton ball of the pure uranium pitchblende. Yet, in the monograph published long before by F. Soddy [2] and in other foreign publications the critical mass of a uranium ball was estimated as 40 tons and its radius as 274 cm. It has recently been established experimentally and is now of common knowledge that a nuclear chain reaction cannot take place in a ball made of natural uranium without moderator, whatever its size.

At that time the RIAN in Leningrad was a leading Russian institute engaged in the uranium processing technology, radium extraction and fabrication of radioactive sources. The RIAN possessed one of the best experimental bases for nuclear physics and radiochemistry created under the guidance of V.I.Vernadskiy and Academician V.G. Khlopin (since 1939). Since the foundation of this institute in 1922 its basic research lines were the properties of radioactive materials and their abundance in the Earth's crust, technology for processing the uranium ore, nuclear physics, and radiochemistry. In the prewar years the first

European cyclotron was built at RIAN, where high-intensity 3.2-MeV proton beams were generated in March–June, 1937.

A technology for extracting radium from uranium ore developed under the guidance of Khlopin made it available in 1930s to supply research facilities with radioactive sources and, after 1932, neutron sources. As noted in the paper *The Atomic Gun* (1937) [3], the 35-ton cyclotron built at RIAN with the help of *Bolshevik* and *Elektrosila* plants was the most powerful in Europe. In Mysovskiy's opinion [3], similar facilities, yet of the lower capacity, were available at that time only in the USA (four cyclotrons). The cyclotron was employed both in physical studies and in the daily production of small (gram-scale) amounts of artificial radioactive isotopes, applied in medicine as a replacement for radium to treat cancer patients. The shortage of radium at that time was extreme. It was on March 1, 1923 that under a decree issued by the Labor and Defense Council radium was recognized as a material controlled by the State Reserves Fund, and this status remained until the late 1970s. In total only about 2 kg of radium was extracted throughout the world by 1937. Radium was primarily used by medical institutions for treating malignant tumors and other diseases [3]. The "radium industry" provided the extraction of milligram amounts of radium from tons of the uranium ore. At that time the "industry" consisted of the RIAN Institute and Bereznikovsk Radium Plant. The founders of this industry were V.G. Khlopin, I.Ya. Bashilov, and A.E. Fersman (Vernadskiy's deputy). In his monograph Academician A.L. Yanshin describes in detail conditions under which the Radium Plant was built [16]: "In July, 1918, the Commission for Natural Labor Forces approved the investments in the study of radioactive minerals in 1918. At the same time funds were allocated to the Chemical Industry Department of the VSNKh (Russian abbreviation for All-Union Council of National Economy) for financing the pilot plant where radium was extracted from the uranium ore." This plant was constructed on the Kama River, near the *Tikhie Gory* pier located close to Bereznyaki. In May, 1920, the uranium ore arrived, which had been mined before 1918, at the Tyuya-Muyun mine in Fergana. As evidenced by Yanshin, on December 1, 1921, the first grams of the Soviet radium bromide were produced. Later, under supervision of Khlopin and Mysovskiy a facility was constructed at RIAN, which extracted radon from radium to supply physicists and physicians operating in numerous research and medical centers of the USSR.

The outstanding scientists of the Soviet Union – I.V. Kurchatov, A.I. Alikhanov, A.P. Vinogradov, M.G. Meshcheryakov, and S.N. Vernov worked on their research probation at the cyclotron and other facilities of the RIAN Institute. The last prewar discovery – spontaneous fission of the uranium nuclei – was made by K.A. Petrzhak and G.N. Flerov (June 1940), workers of the RIAN and LFEI, under the guidance of Kurchatov, who worked at RIAN from 1935 to 1940.

The prewar period was marked by the discovery of two transuranium elements which later on were called neptunium and plutonium. It was found that bombardment of ^{238}U by neutrons could produce ^{239}Pu as a result of the following reaction:

$$^{238}_{92}U + n \rightarrow\,^{239}_{92}U \xrightarrow[23\text{min}]{\beta^-} \,^{239}_{93}Np \xrightarrow[2.35\text{days}]{\beta^-}\,^{239}_{94}Pu.$$

^{239}Pu has the half-life of 24,400 years and converts into ^{235}U via the reaction

$$^{239}_{94}Pu \xrightarrow[24400\,\text{years}]{\alpha}\,^{235}_{92}U.$$

It was in March 1941 that G. Seaborg produced the first microscopic quantity of ^{239}Pu. It was found that plutonium when bombarded with neutrons, much like the isotope ^{235}U, decayed and emitted approximately 3 neutrons. So it became evident that besides ^{235}U, one could use plutonium as a nuclear explosive. However, plutonium could be produced in considerable quantities only in a nuclear reactor. To build the latter it was necessary to have a large amount of uranium.

A Commission on the Uranium Problem, under the Presidium of the USSR Academy of Sciences, was established on July 30, 1940, as a response to V.I. Vernadskiy and V.G. Khlopin's initiative. The Commission objectives were to study uranium properties and the possibilities of utilization of atomic energy. Among the members of the Commission were:

- V.G. Khlopin — Academician, Director of RIAN, Chairman;
- V.I. Vernadskiy — Academician, Director of RIAN (until 1939);
- A.F. Ioffe — Academician, Director of LFEI;
- S.I. Vavilov — Academician, Director of the Institute of Physics, USSR Academy of Sciences (FIAN), Scientific Supervisor of the State Optical Institute;
- A.E. Fersman — Academician, RIAN;
- A.P. Vinogradov — Professor, Laboratory of Geochemical Problems, USSR Academy of Sciences;
- I.V. Kurchatov — Professor, LFEI, RIAN;
- Yu.B. Khariton — Professor, Institute of Chemical Physics (IKhF).

In fact the activity of this commission, which coordinated research works in many institutes, was not covered in detail in publications. Its recommendations and decisions were of crucial value in creating the experimental bases at the

LFEI, UFEI, and other institutes, and, also, in general management of work on nuclear chain reactions. Proposals made by Academicians Vernadskiy, Fersman, and Khlopin were of fundamental importance for solving the uranium problem [17]. In particular, they made the following proposals:

1. The Academy is to organize the prompt development of methods for separating the isotopes of uranium and construction of appropriate facilities. The Academy is to approach the Soviet government for allocating special funds and, also, precious and non-ferrous metals in quantities required.
2. The Academy is to develop a project of super-powerful cyclotron at FIAN.
3. A State Uranium Fund is to be created.

By 1941 the Commission elaborated the following plan of research works:
- investigate a mechanism of separating uranium and thorium (RIAN, LFEI, UFEI);
- define conditions for a nuclear chain reaction in natural uranium (IKhF, LFEI);
- develop methods for separating the isotopes of uranium (RIAN, LFEI, UFEI, and others).

In early 1941 I.V. Kurchatov published a paper *Fission of Heavy Nuclei* where he discussed the mechanism of fission of various heavy nuclei and the possibility of a nuclear chain reaction [18]. Based on the data available on the cross-sections of interactions between neutrons and nuclei and his own calculations he concluded: "...although the possibility of a nuclear chain reaction has now been established in principle, any practical attempts to initiate this reaction in various systems under investigation encounter tremendous difficulties." To illustrate this conclusion he presented a table where the available world reserves of critical materials and the amounts required to implement a nuclear chain reaction were compared. In this paper Kurchatov wrote: "Perhaps the following years will provide us with other ways of solving this problem. Otherwise, only new and high-effective methods for separating the isotopes of uranium and hydrogen will make a chain nuclear reaction possible."

This paper and a number of other publications resulted in a Program proposed by the Uranium Commission, USSR Academy of Sciences. Among others there were some works on prospecting uranium deposits suitable for mining. On October 15, 1940, a resolution was passed charging the Academy with a task to approve the plan of the Uranium Commission and allocate corresponding funds to contractors. The same resolution established a subcommission of the Uranium Commission, headed by A.E. Fersman, responsible for the prospecting, exploration, and development of uranium deposits. The research plan also assumed the construction of new accelerators and other facilities. In February, 1940, first tests of the magnetic unit (75 tons in weight) designed for the LFEI cyclotron lab started at the Leningrad plant

Table 1 Amounts of materials required to perform a nuclear chain reaction [18].

System	Critical material		
	Name	Amounts required, tons	World reserves, tons
Enriched ^{235}U combined with water as moderator	Uranium with double concentration of ^{235}U	0.5	2×10^{-12}
Natural uranium and ^2H	Heavy water	1.5	0.5
Protactinium	Protactinium	≈ 0.2	1×10^{-6}

Elektrosila. The Uranium Commission coordinated research work on radioactive ore and the geochemistry of natural radioactive isotopes. In 1940, experimental regional studies of the radioactivity of water and formation rocks in the area of the Caucasian mineral springs were conducted. Later on, the abundance of uranium in natural waters studied at that time allowed researchers to reveal regularities in the uranium migration and discover potential mining locations.

In September 1939, a Uranium Society was established in Germany and a program for utilization of uranium fission energy was developed. In fact this was the first uranium project in the world. The Kaiser Wilhelm Institute of Physics headed by well-known physicist W. Heisenberg became the scientific core of this project. This institute was subordinate to the Ministry of Armaments. In May 1940, it became common knowledge that Germany had launched a research project aimed at creating atomic weapons. At that time 1200 tons of uranium concentrate previously mined in Congo and destined for the extraction of uranium were captured by the Germans as they invaded Belgium. That was nearly half of the world reserves. Earlier, when Czechoslovakia was occupied in March 1939, the uranium mined in Jachymov and destined for delivery to other countries was also transported to Germany. About the same time an attempt was made to transfer a stock of heavy water from Norway to Germany in order to use it in a nuclear chain reaction, because graphite produced at that time had many unfavorable impurities and could not be used as a moderator in the nuclear pile [19, 20]. All these resources allowed the Germans to successfully advance in constructing a nuclear reactor for production of plutonium and, eventually, in creating nuclear weapons.

In 1940, German scientists discovered that ^{235}U could be used as a nuclear explosive. Immediately a project was started under the leadership of P. Gartec for separating the isotopes of uranium. In July 1940, another German scientist K.-F. von Veiszekker established theoretically that ^{238}U in a nuclear pile should convert into a new element, which had the same properties as ^{235}U and could be used as the nuclear explosive. Thus, Veiszekker independently discovered the element whose existence was later experimentally confirmed by American researchers. This element was called *plutonium*. The first research reactor was constructed under the guidance of W. Heisenberg in December 1940; by that time one of the German companies started the production of metallic uranium. However, they failed to obtain a self-sustaining nuclear chain reaction.

In 1941 participants of the German Uranium Project discovered that the number of neutrons emitted in the reactor by the designed mass of uranium was higher than that of absorbed neutrons. The researchers were close to success, i.e. to putting the reactor into operation. Later W. Heisenberg recalled: "In September 1941 we saw a way opened to us. It was the way to an atomic bomb."

The war speeded up nuclear weapons projects, first in Great Britain, and then in the USA and USSR. In April 1940, G.P. Thompson, one of the outstanding world scientists, took charge of the British Governmental Committee on the Military Use of Atomic Energy. Within a year, on July 15, 1941, Thompson's Committee has completed its report, which concluded that an atomic bomb could be built before the end of the war [20]. Based on this report, a research and development program aimed at creating nuclear weapons in Great Britain was approved in October 1941. Well-known scientists who emigrated to the USA before the war (A. Einstein, L. Szilard, E. Fermi, and others) tried to convince the US Government to develop a nuclear project and outstrip Germany in creating nuclear weapons. On October 11, 1939, A. Einstein wrote a letter to President Roosevelt in which he insisted on the necessity of initiating the development of new weapons. However, a decision to start this project was made by the US Government only on December 6, 1941, following the presentation of Thompson's report [19]. An important factor in making the decision was the Japanese attack on the US military base in Pearl Harbor. Three atomic bombs were manufactured in the USA by the end of the war. This is how a nuclear deterrent evolved, first against Germany and Japan, and then against the USSR.

NOTES

1   The RIAN Institute was created in 1928, and belonged to the Narkomtyazh-mash Commissariat (the same as LFEI). At the beginning of the Uranium Project the institute was tentatively named as Laboratory No. 1. Later it became the Kharkov Physical Engineering Institute (KhFEI).

2   In many publications authors consider the creation of this group as the first stage of nuclear physics research in the USSR. However, one should take into account that one of the RIAN departments was engaged in studying radioactive transmutations since 1922. In the nuclear group created at the LFEI Institute under the guidance of A.F. Ioffe and I.V. Kurchatov (10

researchers in all), G.A. Gamow and L.V. Mysovskiy, the leading Soviet nuclear physicists from RIAN, were engaged as consultants.

3 G.A. Gamow – the outstanding theoretical physicist, the Corresponding Member of the USSR Academy of Sciences since 1932. In 1934 he emigrated to the USA, where he worked at Bohr's laboratory. He is the author of the tunnel effect explaining alpha-decay and, also, of papers in quantum mechanics and nuclear physics. Together with E. Teller he established a selection rule in the theory of beta-decay.

CHAPTER 2

War Period (1941–1945)

From the beginning of the Great Patriotic War research on the utilization of nuclear fission energy in the USSR was suspended. Leading institutes from Leningrad, Moscow, and other cities were evacuated, and many scientists went to the front.

After 1939, when the World War II began, research on uranium nuclear fission was aimed not at maintaining a controlled chain reaction, but at the production of explosive effect, i.e., at the creation of a new nuclear weapon.

By mid-1942 the State Defense Committee (GKO) received information that Germany, possessing a great scientific and engineering potential and large natural uranium reserves, was involved in secret research aimed at the creation of a new powerful weapon — an atomic bomb. In March, 1942, L.P. Beriya, People's Commissar of Domestic Affairs, submitted to Stalin a draft memorandum summarizing the information delivered from London, which described developments in the production of uranium bombs in Great Britain. In this memorandum Beriya recommended that a Scientific Consultative body be established under the GKO Committee authorized to coordinate all domestic research on the uranium problem and familiarize prominent Soviet specialists with the intelligence information about the uranium problem which was at the disposal of the NKVD (People's Comissariat of Domestic Affairs). Among those scientists were P.L. Kapitsa (IFP), D.V. Skobeltsin (FIAN), and A.A. Slutskiy (KhFTI). Evidence of the contribution made by intelligence to the creation of nuclear weapons was reported by A.A. Yatskov, veteran of the KGB intelligence service, in his paper delivered at a seminar which took place in February, 1992, at the Kurchatov Institute of Atomic Energy (IAE) with the participation of researchers from the Natural History Institute of the Russian Academy of

Sciences. In the 1940s Yatskov was a deputy of the Soviet resident in the USA and was responsible for maintaining intelligence communication between Los Alamos (USA) and Laboratory No. 2.[1] The necessity of resuming research on the uranium project in the USSR was emphasized by lieutenant G.N. Flerov, a young scientist, who contributed to the discovery of spontaneous uranium fission in 1940. He spoke at a session of the Academy Small Presidium held in Kazan, to where the Academy institutes (RIAN, LFEI, IKhF, etc.) had been evacuated. The eminent physicists A.F. Ioffe, P.L. Kapitsa, and others, reasoned that the renewal of this research was impossible because it required the diversion of a large amount of human and material resources. Flerov recalled later that he had not let the matter drop. He scanned available foreign physics journals and found out a complete absence of publications on uranium fission and atomic energy utilization by scientists from the USA, Great Britain, and Germany. This meant that nuclear research had gone secret in those countries. Flerov informed the GKO Committee and Stalin about it [21]. Obviously, this information eventually stimulated a decision to renew research on the Uranium Project. In late 1942 the GKO Committee requested the Academy of Sciences to start and coordinate research into the utilization of atomic energy for military purposes. According to the GKO resolution adopted on November 27, 1942, the Narkomtsvetmet Commissariat was charged to arrange the production of uranium from domestic raw materials. I.V. Kurchatov was chosen among other candidates and appointed a scientific director of the nuclear project by a governmental decree [22]. M.G. Pervukhin, Deputy Premier and People's Commissar of the Chemical Industry, was put in charge of supplying necessary materials, instruments and the like for the industry and Academic Institutes and of supervising all domestic work connected with this problem. Usually, it was by his permission that Kurchatov acquainted other scientists with intelligence materials.

By late 1942 it was known that not only ^{235}U nuclei were capable of fission under the action of neutrons but also the nuclei of a new artificial element, plutonium, discovered in 1940–1941, which substantially broadened the field of research.

In 1940–1941 research aimed at discovering elements 93 and 94 was conducted in the US secretly, and these elements were not even named. In 1941 element 94 was referred to as "copper" until a need arose to use real copper in some experiments. For some time plutonium was still referred to as "copper" and copper itself as "genuine copper"... Only in March 1942 element 94 was named plutonium for the first time [23]. In late 1940 element 93, an isotope of neptunium $^{238}_{93}$Np, was discovered during the bombardment of uranium with deuterons. This isotope was produced by the following reaction:

$$^{238}_{92}\mathrm{U} + {}^{2}_{1}\mathrm{H} \rightarrow {}^{238}_{93}\mathrm{Np} + {}^{1}_{0}\mathrm{n},$$

and ^{238}Pu was produced in its decay product:

$$\ce{^{238}_{93}Np} \xrightarrow[\text{2.1 days}]{\beta^-} \ce{^{238}_{94}Pu} \xrightarrow[\text{86.4 years}]{\alpha} .$$

In the spring of 1941, heavier isotopes of neptunium and plutonium were discovered as $\ce{^{238}_{93}Np}$ was bombarded with neutrons:

$$\ce{^{238}_{92}U} + \ce{^{1}_{0}n} \rightarrow \ce{^{239}_{92}U} + \gamma;$$

$$\ce{^{239}_{92}U} \xrightarrow[\text{23.5 min}]{\beta^-} \ce{^{239}_{93}Np} \xrightarrow[\text{2.35 days}]{\beta^-} \ce{^{239}_{94}Pu} \xrightarrow[\text{24000 years}]{\alpha} .$$

About a year and a half later, on December 2, 1942, the first controlled nuclear chain reaction was performed in a natural uranium–graphite system under the guidance of E. Fermi, an eminent scientist of the 20th century. This outstanding experiment confirmed the possibility of producing $\ce{^{239}Pu}$ in large amounts to be used in nuclear weapons. As is seen from his notes to Pervukhin (March 7 and 22, 1943), Kurchatov was informed about these results by the intelligence service. In particular, Kurchatov wrote: " ...there are fragmentary remarks indicating a possible use of $\ce{^{235}U}$ as well as $\ce{^{238}U}$ in a uranium reactor... It is possible that products generated by the combustion of nuclear fuel in a uranium reactor can be used instead of $\ce{^{235}U}$ as a material for a bomb."

On February 15, 1943, the GKO made a decision to establish a research center under the guidance of Kurchatov for the purpose of creating nuclear weapons. In its resolution of April 12, 1943, the Presidium of the USSR Academy of Sciences assigned Laboratory No. 2 a status of independent research organization. In April, 1944, researchers who were earlier enlisted in Laboratory No. 2 and were already working as small research teams accommodated in other Moscow institutes (Seismological Institute, Institute of General and Inorganic Chemistry – IONKh) were transferred to special buildings located in the Pokrovsko-Streshnevo area in the suburbs of Moscow [24]. On August 14, 1943, Kurchatov and a group of his co-workers were moved from LFEI to Laboratory No. 2. This group included A.I. Alikhanov, G.Ya. Shchepkin, P.E. Spivak, V.P. Dzhelepov, L.M. Nemenov, G.N. Flerov, and M.S. Kozodaev [25]. The laboratory staff was reinforced with specialists that were recalled or demobilized from the army. In February 1944 Laboratory No. 2 received the status of research institute. In 1949 it was named LIPAN (Laboratory of Measurement Instruments, Academy of Sciences), and later the

Kurchatov Institute of Atomic Energy. At the present time it is known as the Russian Research Center or simply the Kurchatov Institute.

At that time Kurchatov was charged with a task of great importance – to create atomic weapons. During the war period Laboratory No. 2 had a small staff mostly engaged in constructing an experimental base, drawing uranium and graphite engineering specifications, and performing theoretical computations.

One of the major units of Laboratory No. 2 was Sector 1 headed personally by Kurchatov and engaged in designing a uranium-graphite reactor providing a self-sustaining nuclear chain reaction.

In late 1946, 74 employees of Sector 1 (including 14 researchers and engineers) were engaged in the preparation, installation, and putting into operation of the first Soviet research reactor [26]. Thus, by the end of 1945 Sector 1 and its collaborators completed preliminary research and produced engineering specifications for the allied organizations involved, primarily concerned with the required amounts and quality of uranium and graphite. As a result it took Laboratory 2 only one year (1946) to construct the first nuclear reactor which was started at 18.00 on December 25, 1946 [26, 27]. The building of the reactor, the fabrication of graphite and uranium items, and the manufacturing of equipment were accomplished under the direct control of the Special Committee and the First Chief Directorate (PGU), both established in August 1945.

In mid-1943 Pervukhin appointed V.V. Goncharov, the director of an experimental plant in Baku, as an assistant of Kurchatov. Goncharov was primarily put in charge of arranging at plants in other industries the production of pure graphite and metallic uranium, the basic materials to be used in the active zone of a nuclear reactor. A review [24] describes in detail numerous problems encountered in creating the domestic atomic industry. Goncharov later recalled that in 1943–1944 the researchers of Laboratory No. 2 carried out theoretical and experimental studies and revised the prewar calculations of Yu.B. Khariton and Ya.B. Zeldovich concerning the possibility of maintaining a nuclear chain reaction. It was found that ^{238}U fission was initiated only by fast neutrons, whereas slow neutrons were absorbed without fission.[2] This prevented a chain reaction in a homogeneous uranium-graphite mixture because fast neutrons were slowed down in the presence of graphite. Calculations made by I.I. Gurevich and I.Ya. Pomeranchuk proved the necessity of building a heterogeneous pile where graphite bricks and bars of natural uranium were arranged at a fixed spacing inside the reactor.

Engineering specifications for graphite bricks ($100 \times 100 \times 600$ mm in size) were drawn by March 1945. According to these specifications the average content of ash and boron in pure graphite should be no more than $3 \times 10^{-3}\%$ and $2.45 \times 10^{-5}\%$, respectively.

The State Research Institute of the rare metals industry (GIREDMET Institute), People's Commissariat of Non-ferrous Metals, worked out a technology for producing extraordinarily pure uranium metal to the specification of Laboratory No. 2. In December 1944 the first 1-kg bar of pure metal uranium was produced by the GIREDMET technology (by N.P. Sazhin with participation of Z.V. Ershova). A decree of the State Defense Committee of December 4, 1944, set up an Institute of Special Metals (later on named NII-9) and an experimental Plant No. 5 in the NKVD system. Professor V.B. Shevchenko, a colonel-engineer, was appointed as director of the institute and plant by a decree of January 3, 1945. This institute and Laboratory No. 2 were accommodated in the same unfinished premises of the All-Union Institute of Experimental Medicine (VIEM) located in Pokrovskoe-Streshnevo in the Moscow suburb. Only at the end of 1945, was a decision taken to arrange the production of metal uranium and uranium articles at Plant No. 12 which belonged to the People's Commissariat of Munitions and was situated in Elektrostal City. Thus, by the end of the war only preliminary research in the physical basics of reactor F-1 and the preparation for supplying Laboratory No. 2 with basic materials were completed. Until 1945, researches in the other sectors of Laboratory No. 2 were performed by a limited manpower. By April 25, 1944, the overall number of workers involved in all research lines of Laboratory No. 2 totaled about 80 including 25 scientific researchers. However, many other institutes such as RIAN, LFEI, IKhF, and IONKh, as well as the GIREDMET institute, Plant No. 12, the Moscow electrode factory, and a number of other enterprises participated in the solution of the uranium problem at that time. Until mid-1945, the researches dealing with creation of an atomic bomb and atomic industry as a whole were rather limited. Despite the intensive labor and important results achieved by that time, these small teams were incapable of producing sufficient scientific and material resources for the creation of an atomic industry. That was obviously not only because of the war. V.V. Goncharov recalls: "In my opinion, this situation can be explained by the following reasons. The Soviet government did not give due credit to the information that work was being done in Germany and the USA for creating superpowerful weapons. However, to be on the safe side an institute was set up to deal with that problem. It was only after the first US bomb was exploded in July 1945 and two bombs were dropped on Hirosima and Nagasaki in August 1945 that they believed that the atomic bomb was a reality and it was not a fruit of physicists' fantasy." These conclusions made by V.V. Goncharov were supported by N.M. Sinev who later became Chief Designer of the special design bureau for devising the technology and equipment intended to produce enriched uranium for nuclear weapons. In his review published in 1991 [28] he described the history of the industrial technology and production of highly enriched uranium in the former USSR. He notes that in May 1945, I.K. Kikoin and the staff of his small laboratory were transferred from Sverdlovsk (Uralian Branch of the USSR Academy of Sciences) to Laboratory No. 2. In November 1945, they started research into various methods for separating uranium isotopes.

Actually, extensive research started only after they received the news of the Hiroshima and Nagasaki events. In September 1945 a discussion was held on the research and practical results of Laboratory No. 2 in the utilization of atomic energy. One of the papers *On the Research of Producing Plutonium in Uranium-Graphite Reactors Cooled with Light and Heavy Water Agents* was presented by I.V. Kurchatov, G.N. Flerov, and A.I. Alikhanov.

Thus, until the conclusion of the Great Patriotic War very small human resources were involved in the uranium project. Full-scale research into the creation of nuclear weapons was initiated after the explosion of three US ^{235}U and ^{239}Pu bombs. Practical work was speeded up by intelligence information provided by Klaus Fuchs, a British scientist who had worked on problems related to the creation of nuclear weapons since 1941, first in Britain and later in the USA [29]. When he learned that the project for creation of nuclear weapons was kept secret from the ally, the USSR, Fuchs reckoned this situation to be wrong. Residents of the Soviet intelligence made a contact with Fuchs and regularly transmitted information to Moscow. From 1943 to 1946 Klaus Fuchs and other agents delivered information about the creation of nuclear weapons including data about a nuclear reactor used in production of plutonium and a detailed drawing of the first US nuclear charge. Yet, in spite of the intelligence information on the foreign projects aimed at the creation of nuclear weapons, including the information provided by Fuchs, the initiative of G.N. Flerov, and a letter written by Academician N.N. Semenov in 1940 where he emphasized the necessity of starting up work for the creation of nuclear weapons, large-scale research was not started until the end of the war. In this context it is pertinent to cite a report presented by E. Teller at the 1962 All-American conference on the problems of management of complex programs including a program for the creation of atomic and hydrogen bombs [30]: "We had a choice of four possible ways to produce fission materials. The production of fission materials is the most labor-consuming stage in creating an atomic bomb. Having succeeded in solving this problem, any nation can within a few months possess a bomb." Of particular difficulty in any atomic project was the necessity to carry out large amounts of design, engineering development, and mastering the production of fission materials. Only one country, the USA, possessed a powerful economy that was capable of mobilizing all required materials and human resources to do this work during the war. This conclusion is supported by the intelligence information that $2 billions were spent and about 130 thousand workers were recruited to set up the Tennessee Atomic Center where a plant for producing ^{235}U was constructed. Besides this center many other facilities were established.

What are the four possible ways of producing nuclear explosives, which in Teller's opinion were much more complicated than the designing and construction of an atomic bomb itself? Those ways were already known at that time. At the very beginning of the practical implementation of the US atomic project, a special report was submitted to the President of the USA on July 17, 1942, on three methods for producing ^{235}U and one method for producing ^{239}Pu [30]. Those four methods are as follows: electromagnetic, gas diffusion, and centrifuge separation of uranium isotopes, and production of ^{239}Pu by a controlled self-sustaining nuclear chain reaction.

The problem was to choose a way that would be cheap and easy to implement. In his report E. Teller mentioned that a few kilograms of ^{235}U or ^{239}Pu had an explosive effect equal to a few thousand tons of conventional explosives. He emphasized that such a bomb could be fabricated and exploded at any time desired, that to produce this explosive one could safely construct large industrial facilities, and that given the necessary resources the program should be realized within a short term to be of military significance.

F.D. Roosevelt issued an order to start up work on the atomic bomb immediately. The project was placed under the control of a newly established subunit of the Army Corps of Engineers. All necessary research personnel, premises, laboratories, production units, and security and safety corps were governed by Colonel L.R. Groves of the Army Corps of Engineers. The project was officially authorized on August 13, 1942, and named the Manhattan Project. The Manhattan Project included the following important phases:

1. The physical start-up of the first world research nuclear reactor was performed on December 2, 1942, under the guidance of E. Fermi (the reactor was constructed under the stands of the Chicago stadium). A chain reaction began at 3.25 p.m. local time. The "atomic fire" was allowed to burn for 28 minutes. This experiment proved the feasibility of creating nuclear weapons. The critical mass of the reactor was as much as 46 tons of uranium placed in the 385 tons of pure graphite. Holes were provided inside the graphite bricks to accommodate uranium bars. To control the nuclear chain reaction, a few canals were provided in the graphite reactor assembly to host adjustable bronze rods coated with cadmium (strong absorber of thermal neutrons).

2. A special-purpose plant and a large number of experimental facilities were built in the Tennessee Valley, which were designed for producing enriched ^{235}U from uranium ore (Plant Y-12). This plant gave birth to a new city, Oak Ridge, with a population of 79,000 residents. Here, in 1943 the large Clinton (later named Oak Ridge) National Laboratory was set up to develop electromagnetic and gas-diffusion technologies for producing enriched ^{235}U, study transuranium elements, reprocess the wasted fuel, etc.

3. A new city, Hanford, and a large plutonium production center (Hanford Atomic Works) were built on the south bank of the Columbia River, Washington. Immediately after the Chicago reactor was put into operation, uranium-graphite water-cooled reactors, radiochemical plants, and other facilities designed to produce plutonium for nuclear weapons began to be built there. L. Groves recalled in his paper on the program

of atomic bomb creation [30]: "At first, we did not know what form plutonium would have if we managed to get it at normal temperature and pressure: solid, liquid, or gaseous... We completely depended on the theory and could get scarce practical data with the aid of microscopic radiochemistry." Later, on the basis of the Hanford Center the Pacific Northwest laboratory and Hanford laboratory of engineering and technical development were established (both in Richland City).

4. The construction of the Los Alamos Research center for developing a design for an atomic bomb was launched in New Mexico.

The Manhattan Project consisted of several subprojects headed by outstanding physicists:

- electromagnetic separation of uranium isotopes – E. Lawrence, Head of the Radiation laboratory, University of California, Berkeley;
- gas-diffusion separation of uranium isotopes – H. Urey;
- creation of nuclear reactors at Hanford and the production of plutonium for the atomic bomb – A. Compton, E. Fermi, and Eu. Wigner;
- thermo-diffusion method of separating uranium isotopes – Ph. Abelson.

J.R. Oppenheimer headed the Los Alamos National Laboratory from 1943.

It should be noted that before the Manhattan Project was launched all laboratory studies as well as experiments at experimental facilities were carried out under the control of the Science Research and Development Department. L. Groves recalls [30] that it was only in the middle of 1942 that a decision was taken to transfer the guidance and control of the project to the Manhattan Army Corps of Engineers.

The final result of the US nuclear weapons project is well known:

- at 5.30 a.m. on July 16, 1945, the first atomic bomb fabricated of plutonium was successfully tested at the Los Alamos test site. The explosive yield was equivalent to 20 kilotons TNT [19];
- at 8.15 a.m. on August 6, 1945, the second atomic bomb fabricated of a highly enriched uranium-235 was exploded over the Japanese city of Hiroshima;
- at 11.02 a.m. on August 9, 1945, the third atomic bomb in which a plutonium nuclear explosive was used was exploded over the Japanese city of Nagasaki.

In the opinion of many US scientists and politicians the atomic bomb program was a success thanks to the enthusiasm and hard work of all its participants. They knew the final goal they worked for.

NOTES

1 V.B. Barkovskiy, another veteran of the intelligence service, who had worked in Great Britain, spoke at seminar held at Kurchatov Institute on October 26, 1994.

2 Under certain conditions slowing neutrons can be intensively absorbed by ^{238}U (resonance absorption).

CHAPTER 3

Infancy of Nuclear Weapons Creation in the USSR

Japan, a former ally of Germany, still carried on military operations even after Germany had surrendered. The atomic explosions over Hiroshima and Nagasaki resulted in its capitulation, and also demonstrated how atomic weapons can be manipulated by politicians. Academician A.D. Sakharov wrote in his memoirs [31]: "It was the irony of fate that in 1945, Teller and Szillard recommended just a demonstration of the atomic bomb rather than its military use, whereas Oppenheimer reasoned that this problem should be left to the military and politicians for solution (Teller recalls that he was too easy to change his mind)." Subsequently, the attitude of the American creators of nuclear weapons to the use of their creation changed to the opposite.

The year of 1945 was a turning point for the situation in the Soviet Union. All organizational operations aimed at the preparation of the industrial base for manufacturing nuclear weapons were concluded. The coordination role of Laboratory No. 2 increased considerably after the establishment of two special governmental organizations. In fact, a perfectly coordinated project for creating a new industry was launched, which resulted in the production of Soviet nuclear weapons. By a decree No. 9877 issued by the GKO Committee (State Defense Committee) on the 20th of August, 1945, a Special Committee was set up which was furnished with special and extraordinary powers for solving any problems related to the Uranium project. The Committee consisted of:

Beriya — Chairman;
Pervukhin — Deputy Chairman of the USSR Sovnarkom Committee;
Voznesenskiy — Chairman of the USSR Gosplan Committee;
Malenkov — Secretary of the KPSS Central Committee;
Vannikov — People's Commissar of Munitions;

31

Makhnev — Secretary of the Special Committee;

Kapitsa — Academician, Director of IFP, AN SSSR;

Zavenyagin — Deputy People's Commissar of Domestic Affairs;

Kurchatov — Head of Laboratory No. 2, AN SSSR, scientific supervisor of the problem.

The same GKO decree established a Technical Council of the Special Committee, and on December 27 an Engineering Technical Council was set up. The Special Committee and its Technical and Engineering Technical Councils started active work aimed at the creation of the atomic industry. V.M. Molotov recalled that Kurchatov paid tribute to the intelligence information about the implementation of nuclear weapons projects that came from Great Britain and the USA in 1943. This information and the fact that Stalin and Molotov were aware of the statement made by President Truman at the Potsdam Conference in July 1945, in which he mentioned that the USA possessed an atomic bomb, forced the Soviet Government to take extraordinary measures: sites were chosen in the Ural region for a plutonium center and a plant intended for the production of enriched ^{235}U. Molotov recalled [22]: " ...though Truman tried to dumbfound us by his information about the creation of unusual superdestructive weapons in the USA, it was obvious they had not enough weapons (atomic bombs) to unleash a war against the USSR. In our country research in this line was only getting started at that time."

D.D. Eisenhower recalled that the problem of using atomic bombs to conclude the war against Japan as soon as possible was discussed at the Potsdam Conference in July, 1945 [32]: " ...a decision was taken that the plan of using atomic bombs against Japan would be carried out if the latter did not proclaim a prompt surrender in accordance with the requirements delivered to the Japanese government from Potsdam." Thus, the USSR government had at its disposal not only an intelligence message about the US nuclear project but also official information about the military use of atomic weapons. In 1946 the Special Committee was put in charge of the construction of an industrial nuclear reactor for producing plutonium concurrently with the preparation for launching the first nuclear research reactor F-1.

The construction site for the first industrial reactor intended to produce weapon plutonium was chosen and surveyed in late 1945 [33]. The first group of builders was sent there immediately. Ya.D. Rappoport was appointed director of the building project, and V.A. Saprykin a chief engineer. In the summer of 1946 a foundation pit started to be excavated for the first industrial reactor at a site 70 km north of Chelyabinsk. On April 17, 1946, P.T. Bystrov was appointed director of the center under construction.

In 1945 a number of other important decisions were taken:

- On August 3, 1945 the First Chief Directorate (PGU) was established under the Sovnarkom Committee to manage the creation of atomic industry and coordinate all related scientific, technological, and engineering projects. Besides Laboratory No. 2, the following organizations were assigned to the PGU control: a factory of munitions (Plant No. 12, Elektrostal), State All-Union Design Institute (GSPI-11, Leningrad), machine-building plant (Plant No. 48, Moscow), uranium ore mining center (Center No. 6, Tadjikistan), and the NII-9 Institute. The NII-9 Institute was put in charge with all operations with uranium including the metallurgy of uranium and plutonium and their alloys, and the production of articles from them;

- In December 1945 two special experimental design bureaus (OKB) were set up in Leningrad, one at the Kirov Plant (LKZ Plant), the other at the Elektrosila Plant, for the purpose of designing equipment for the gaseous diffusion and electromagnetic production of ^{235}U;

- On December 1, 1945, a decree was adopted to construct a gaseous diffusion plant near the Verkh-Neivinskiy settlement, Middle Ural, for enriched ^{235}U production;

- On December 1, 1945, Laboratory No. 3 was set up to carry out research on creating heavy water reactors using natural uranium;

- In 1946 a site was chosen in the vicinity of Arzamas (Sarov City) for building premises to accommodate a Laboratory No. 2 branch, a research center for designing nuclear weapons (future VNIIEF Institute) [34];

- A plan was proposed to establish a large research center near Moscow for implementing an atomic program of nuclear power industry. The facility was built near the Obninskoe railway station, at the 110th kilometer of the Moscow–Kiev railway [35]. Later the center was transformed to a Physics and Power Institute (FEI).

For the purposes of geological uranium exploration and mining, plutonium and ^{235}U production, and nuclear weapons construction, the Special Committee under the State Defence Committee and the PGU recruited numerous teams of workers from other branches of the industry.

On April 9, 1946, the Council of Ministers[1] approved the PGU departments, sections, and central administration. Thus, a complex program was initiated which perfectly coordinated the activities of numerous ministries, research institutes, and laboratories involved. This program was at that time a top-priority program in the USSR. P.Ya. Antropov was put in charge of uranium exploration and mining. E.P. Slavskiy was in charge of graphite production for reactor F-1 and an industrial reactor intended to produce plutonium for nuclear weapons. A.P. Zavenyagin and A.N. Komarovskiy were responsible for the rapid establishment of new enterprises, research institutes, and secret towns and settlements required for the atomic industry. V.S. Emeliyanov was appointed chief of the PGU research department. Together with I.V. Kurchatov, scientific director of the project, he coordinated and controlled the activities of all research institutes and design bureaus involved.

The staff and administration of the PGU and NKVD Department No. 9 were later accommodated at 8a Novo-Ryazanskaya street, not far from the Kazansky railway terminal. All of the orders of the researchers, designers, and builders were forwarded via the PGU and Special Committee to appropriate organizations to be fulfilled as top-priority. A decree of August 20, 1945, declared that all jobs done under the PGU or by enterprises of other branches to its order were to be controlled by the Special Committee: "No organizations, institutions or persons can interfere without special authorization by the GKO Committee with the activity of the PGU, its enterprises, and institutions or ask for information about its work or jobs being fulfilled to the PGU order." The PGU Committee was authorized to recruit experts from any industry for the atomic industry project.

Front of the building where the First Chief Directorate (PGU) was accommodated.

The PGU included departments for planning, financing, and controlling all of the projects that were implemented by the enterprises, building sites, and institutes governed by the PGU, and also by the organizations engaged from other branches of the industry or institutes of the Academy. There were also departments of personnel, capital construction, and others.

Later, in addition to the specialized departments of the PGU, a new department was added for the separation of uranium isotopes. It was headed by A.M. Petrosyants, a machine-building engineer, major-general, who earlier worked for the Special Committee [28]. The department for uranium mining and production of uranium slugs was headed by S.E. Egorov and N.F. Kvaskov.

A Scientific and Technological Council of the PGU (NTS Council) was set up on April 9, 1946, instead of the previously existing Technical and Engineering-Technical councils under the Special Committee. The NTS Council was chaired by B.L. Vannikov, director of the PGU. The council comprised five sections. They were headed by M.G. Pervukhin[2] (Section 1 − Nuclear reactors), V.A. Malyshev[2] (Section 2 − Diffusion enrichment of uranium), I.G. Kabanov and D.V. Efremov (Section 3 − Electromagnetic separation of uranium isotopes), V.S. Emeliyanov (Section 4 − Metallurgy and chemistry), and V.V. Parin and G.M. Frank, officers from the Public Health Ministry (Section 5 − Medical and sanitary control). I.V. Kurchatov and V.G. Khlopin were appointed scientific heads of the corresponding lines in the atomic research. On November 29, 1947, the USSR Council of Ministers approved the new managerial staff of the Scientific-Technological Council: V.A. Malyshev, I.T. Tevosyan, A.P. Zavenyagin, V.S. Emeliyanov, A.I. Alikhanov, A.P. Aleksandrov, I.K. Kikoin, N.N. Semenov, S.L. Sobolev, I.E. Starik, V.B. Shevchenko, B.S. Pozdnyakov, and some other scientists. Pervukhin was appointed Premier Deputy Director of the PGU; from 1947 to 1949 he acted as Chairman of the NTS Council. Kurchatov was appointed Deputy Chairman of the NTS Council, and in 1949 as its permanent Chairman. Pozdnyakov was appointed Scientific Secretary of the NTS Council; until then he acted as Chief of the Technical Department of the Narkomtyazhprom Commissariat (People's Committee of Heavy Industry).

The rearrangement of the Uranium Project management strengthened the role and raised the responsibility of Laboratory No. 2 and Kurchatov, scientific director of the atomic problem. Work done at Laboratory No. 2, the research institutes, design bureaus, plants, and building sites, as well as the activities of partners from other branches of the industry, was put under a severe, centralized state control. Progress reports on the projects assigned by the PGU to subordinated branches were regularly presented by top officials of various ministries at the sessions of the Special Committee. The PGU administration, NTS Council and its sections regularly discussed the progress of particular phases of the atomic project. The first presentation of this kind took place in September, 1945 at the Technical Council of the Special Committee. The following topics proposed by the scientists of Laboratory No. 2 and other research institutes were discussed:

- plutonium production in uranium-graphite reactors cooled by ordinary and heavy water. Speakers: I.V. Kurchatov, A.I. Alikhanov, and G.N. Flerov (September 5);
- progress of the research into enriched uranium production by the gas-diffusion method. Speakers: I.K. Kikoin and P.L. Kapitsa (September 6);
- uranium enrichment by the electromagnetic method. Speakers: L.A. Artsimovich and A.F. Ioffe (September 10).

Particular goals were formulated in detail, and scientific managers were appointed to supervise specific projects of the atomic problem. On December 16, 1946, by its decree No. 2697–1113 the USSR Council of Ministers set up a Scientific Council under the President of the Academy to supervise all research projects dealing with the atomic nucleus and other scientific projects in line with the PGU activity. This Council allocated research problems to various departmental institutes and controlled the fulfillment of specific plans together with the top officials of the ministries engaged in the execution of Program No. 1.

In late 1945 the Special Committee took a decision to speed up the implementation of the atomic project. In addition to scientists, designers, technologists, and production workers were recruited. In late December the USSR government issued an order launching all enterprises of the atomic industry into immediate operation.

Due attention was given to the problems of radiation effects on human beings and the environment. The Public Health Ministry of the USSR was put in charge of measures providing radiation safety [11]. A.I. Burnazyan, Deputy Minister of Public Health, was appointed the head of special departments in the Ministry and organized similar departments at the institutes and enterprises involved. A system for sanitary supervision and medical control was organized. In 1946, a radiation laboratory (later the Biophysics Institute of the Ministry) was set up. Professor G.M. Frank was appointed its head. The laboratory was in charge of developing the basics of radiation safety and dosimetric control. In the same year, in accordance with a governmental decree on developing research into radiation effects on flora, a biophysical laboratory (headed by Professor V.M. Kluchnikov) was set up in the Timiryazev Academy of Agriculture.

F-1 Reactor Creation

Invention of the first nuclear reactor in the USSR, and in Europe, was first mentioned by V.S. Fursov in July 1955, in his paper at the session of the USSR Academy of Sciences devoted to the peaceful use of atomic energy.

By 1946 the research institutes and enterprises of the Mintsvetmet (non-ferrous metals) and other ministries produced high-purity graphite and uranium in conformity with the strict specifications issued by Laboratory No. 2. For instance, requirements to graphite purity prescribed no more than a few millionth fractions of boron, a strong absorber of thermal neutrons. V.V. Goncharov recalls that the requirements formulated by Laboratory No. 2 were taken by the workers of the plant to be impracticable. They reasoned that it was impossible to manufacture graphite bricks with a purity much higher than that of diamond (or natural carbon). A technology to be used at the plant for producing graphite electrodes was fully changed by the joint efforts of the plant workers and researchers from Laboratory No. 2. Through the use of a low-ash raw material and a number of innovations in the thermal and gas refining processes they managed to devise an advanced technological process for producing graphite of the required purity.

Z.V. Ershova, who at that time headed the Uranium Laboratory at the GIREDMET Institute, recalls that many difficulties had to be overcome in producing metal uranium [11]. The graphite and uranium delivered from the industry to Laboratory No. 2 were substantially variable in properties and purity, so they had to solve a problem of using different-grade graphite and uranium in the first reactor. V.S. Fursov noted that they had to consider the construction of the first uranium-graphite reactor as a decisive experiment to measure the nuclear parameters of a uranium-graphite assembly rather than as a project of assured success. German physicists failed in a similar project in 1941–1942: the neutron absorption cross-section of graphite they measured brought them to the wrong conclusion that they could not use carbon as a neutron moderator in a thermal neutron reactor. So from the outset they abandoned any attempts to build a uranium-graphite reactor.

Experiments aimed at determining the physical characteristics of basic materials for a uranium-graphite assembly started at Laboratory No. 2 in 1944 [26]. By that time researchers managed to get 3.5 metric tons of Acheson graphite and about 220 kg of high-purity uranium oxide. With the use of a technique devised at Laboratory No. 2 they found the effective neutron absorption cross-section of graphite to be equal to $(5\pm3)\times10^{-27}$ cm^2. They also measured the total cross-section effective for the interaction of uranium with slow neutrons and established that 1 kg of uranium spontaneously emitted 24 ± 7 neutrons per second.

It was only in the second half of 1946 that small batches of uranium and graphite started to be supplied to Laboratory No. 2. By November 1946 more than 24 tons of uranium and about 300 tons of graphite were delivered. As uranium and graphite batches arrived, they were used by 30 workers (masons) supervised by A.A. Zhuravlev to build preliminary models gradually increasing their sizes to attain the critical size of a reactor. Special start-up equipment was manufactured to monitor neutron flux. The growth of a neutron flux in the center of the model assembly was simultaneously measured using the activity of indium foil which was placed inside the assembly and was activated as it absorbed neutrons produced during ^{235}U fission.

To create a system providing control of the nuclear chain reaction one should know its effect on the explosion danger, the amount of delayed neutrons generated in the chain uranium fission, thermal extension parameters, the appearance of slag, and the like. I.V. Kurchatov and I.S. Panasyuk [26] wrote: "After the publication of Smyth's book [36] it was possible to determine the effect of delayed neutrons on reactor kinetics: Smyth reported the delay periods and percentages of 4 groups of delayed neutrons. Using these data M.S. Kozodaev found that owing to delayed neutrons a reactor with a supercritical value below 7×10^{-3} would never explode, its kinetics in that case depending mostly on the period of delayed neutrons."

In 1945 it was established that a uranium-graphite reactor with a supercritical value of $(2-3) \times 10^{-3}$ can be controlled using one neutron absorbing cadmium or boron rod, 3-4 cm in diameter, introduced in its core. A system with a supercritical value of 2×10^{-2} needs 8 or 10 rods spaced far enough to avoid the overlapping of their disturbing zones (each within a range of about 1 m). Kurchatov and Panasyuk [26] wrote: "All of the control rods must be kept ready to be immediately loaded into the reactor in the case of a random growth of supercriticality above a dangerous threshold." In a steady operation mode the reactor supercriticality is absent, $K_{eff} = 1$, the power output is permanent, and the reactor is controlled by 1 or 2 rods with the other rods kept standby.

In the experiments conducted with the F-1 reactor it was found that 30-40-mm variations in the diameter of the uranium metal slugs weakly affected the neutron multiplication factor. In order to accommodate the uranium slugs more than 30,000 holes were drilled in the $100 \times 100 \times 600$-mm graphite bricks spaced 200 mm apart. In the final design of the F-1 reactor a square grid with a spacing of 200 mm was used.[3] Uranium slugs of different diameters and rectangular and spherical uranium oxide pellets were placed at the grid nodes. Overall 45.07 metric tons of uranium and 400 metric tons of graphite were loaded in the F-1 reactor (Table 2). The reactor started to operate at a somewhat lower uranium charge (≤ 45 tons).

The historic start-up of a reactor with a specific arrangement of a graphite-uranium assembly in a building with an underground laboratory constructed at the site of Laboratory No. 2 (Figs 1–3) is described in detail in [26], [27]. The goal of the first phase − a self-sustaining nuclear chain reaction − was achieved at 6.00 p.m. on December 25, 1946. The first start-up tests of the F-1 reactor conducted under the guidance of Kurchatov were attended by N.I. Pavlov, an authorized representative of the Council of Ministers, and the researchers of Sector 1 I.S. Panasyuk, E.N. Babulevich, B.G. Dubovskiy, I.F. Zhezherun, A.A. Zhuravlev, N.V. Makarov, K.N. Shlyagin, and also the laboratory assistants A.K. Kondratiev and R.S. Silakov [27].

Table 2 Amounts of uranium metal and uranium oxide loaded into the first nuclear reactor.

Uranium article	Mass, kg	Quantity	Total weight, tons
Uranium oxide − briquets, 49×58×67 mm in sizes	0.88	3143	2.77
Uranium oxide − spheres, 80 mm in diameter	1.18	7473	8.8
Uranium metal − slugs, 32 mm in diameter, 100 mm in length	1.4	2503	3.5
Uranium metal − slugs, 35 mm in diameter, 100 mm in length	1.7	17253	30.0
Total	–	30642	45.07

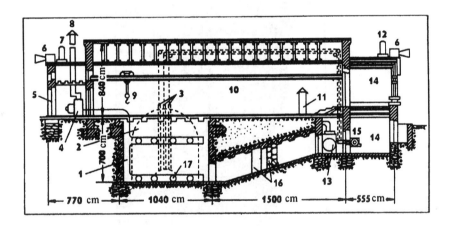

Fig. 1 Longitudinal sectional view of the reactor building in the first years of its operation [27]: 1 – shaft, 10×10×7 m; 2 – reactor; 3 – control and emergency rods; 4, 8, 17 – exhaust air ventilation system; 5 – gate of the building; 6, 7, 12 – sound and light radiation alarm system around the building; 9 – crane; 10 – main hall; 11, 13 – forced ventilation system; 14 – underground laboratory; 15 – winch for remote manual rod control; 16 – gangway from underground laboratory to the shaft.

Although the reactor had no forced cooling, owing to the high thermal capacity of the system they managed to operate the reactor at up to 3.89 megawatts power production within a few minute intervals. The reactor supercriticality ($K_{eff} > 1$) $K_{eff} - 1 = 2 \times 10^{-3}$ was achieved by removing the control rods; the system then self-compensated, and its power output dropped rapidly owing to the heating of uranium. The overall energy production within 30 minutes was 540 kW · hr. During these powerful tests the F-1 reactor pro-

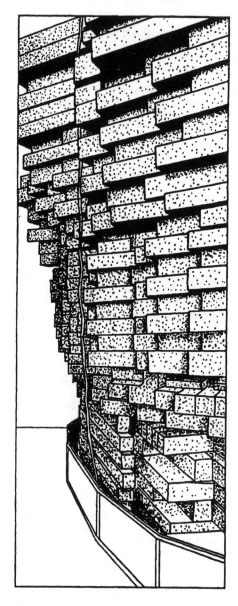

Fig. 2 Shown in black at the foot of the graphite assembly is a metallic barrier protecting the lower graphite layers from accidental water at the shaft bottom [37].

duced the maximum amount of plutonium accumulated in the uranium slugs (milligrams of plutonium per 45 tons of uranium). The F-1 reactor active zone assembled from graphite bricks and uranium slugs, was a sphere, 6 m in dia-

Fig. 3 Assembly of the first nuclear reactor at the supercritical state ($K_{eff} > 1$) [27]: 1-3 – vertical channels of 55-mm diameter; 1′ – cadmium control rod of 50-mm diameter; 2′, 3′ – cadmium emergency rods (50 mm); 4 – hoisting apparatus of a bridge crane; 5 – boxes with uranium slugs; 6 – loudspeaker from a BF_3 start-up panel operator; 7 – ventilation system.

meter. The neutron reflector around it was 8 m thick. There were 3 vertical channels for inserting control and emergency rods (SUZ system) and 6 horizontal experimental channels. Because the reactor had a negative temperature coefficient of reactivity, its power output was spontaneously dropping from 3.89 megawatts to considerably lower values (Fig. 4). Because of slow cooling of the graphite, within 30 minutes after a powerful start-up the SUZ control rods were removed to stop a controlled chain reaction. Based on the analysis of a few F-1 start-ups temperature coefficients of uranium and graphite were determined. The reactor proved to be a self-regulating system providing safe operation.

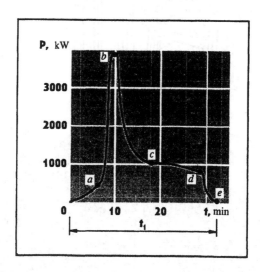

Fig. 4 Typical plot showing variation of power with time at high-power operation (SUZ control rods were removed) [27]: ab – rapid power build-up after reactivity release 2×10^{-3}; b – point of reactor self-compensation ($K_{eff} = 1$); bc – rapid power decline by uranium heating; cd – slow power decline by graphite heating; de – decline after SUZ rod removal.

As noted above, the F-1 reactor had neither cooling system nor biological protection system against radiation. The ventilation system pumped air to the reactor shaft at a capacity of about 7,000 m^3/h and ensured radiation safety only when the reactor power was below 10 kilowatts. At higher power, in the so-called high-power start-up regime, radioactive gases liberated from the uranium reached the adjacent premises making inadmissible the presence of servicing personnel at the reactor. During the high-power start-up tests of the reactor when its power was as high as 100 kilowatts, and even 4 megawatts for short periods, a remote control was used: the reactor was controlled from the main building of Laboratory No. 2 located at about 1.5 km from it. "Prior to each start-up the remote control system was thoroughly examined and the positions of the cadmium rods checked. Immediately after and especially during the high-power operation of the reactor no one was allowed to approach the reactor building because of mortal radiation" [26]. When the reactor operated at 500–1000 kilowatts, the radiation background in the main building of Laboratory No. 2 increased tens times, and the biologically dangerous area around the reactor building was within a range of 200 m. "To protect this area during the high-power tests, the area was guarded by specially appointed watchmen" [26].

The radiation situation during the start-up of the reactor was controlled with the aid of remote dosimeters and dosimeters measuring the air concentration of radioactive gas, both designed at Laboratory No. 2. Researchers from the Radiation Laboratory designed individual integrated dosimeters consisting of thimble-shaped ionization chambers and film detectors. Kurchatov and Panasyuk wrote [26]: "With the aid of these devices we accomplished biological safety control and carried out biological experiments with animals. None of those who serviced the first Soviet atomic reactor were subject to a serious radiation injury. Those few who were exposed were immediately dismissed and sent to a sanatorium for recreation."

Besides testing a controlled nuclear chain reaction and producing plutonium in milligram quantities, in 1947 experiments were made using special neutron-absorbing materials placed into the F-1 experimental channels to produce artificial radioactive elements. These neutron absorbers prevented a critical state ($K_{eff} = 1$) without adding extra uranium.

All necessary experiments were made using the F-1 reactor to determine the dimensions and physical parameters of an industrial uranium-graphite reactor and to prove its feasibility. In these experiments uranium slugs, canned in aluminum or its alloys, were employed. Technological channels (TC) for inserting uranium slugs were also made of aluminum. The slug–channel clearance was to be filled with water used for heat removal. It was necessary to optimize grid spacing for the future industrial reactor active zone and to determine an optimum clearance for the passage of a heat transfer medium (water) through the channels. Water and aluminum were additional neutron absorbers. To determine the critical parameters of the reactor, the grid spacing was modified from 15 to 25 cm, and various diameters of uranium slugs were tested (30, 35, and 40 mm). The clearance between the slug jacket and the aluminum tube was filled with paraffin, comparable with water in neutron absorption. Experiments with 30 different grids revealed that the amounts of uranium and graphite had to be increased 3–4 times to achieve a controlled nuclear chain reaction. Theoretical estimates of a future industrial reactor dimensions were based on equations derived by I.Ya. Pomeranchuk in 1944 for determining the critical mass of a reactor. V.S. Fursov who was later appointed Deputy Science Director of the uranium-graphite reactors project proposed a formula which allowed one to calculate the effective radius of a reactor, i.e., the required amount of uranium and the dimensions of a graphite assembly, based on a difference between the K_{eff} value of the F-1 reactor and the K_{eff} data obtained from the tested grids. The water-filled slug-tube annulus was modified from 2 to 4 mm.

Experiments conducted on the first European research reactor helped determine the optimum characteristics of an industrial reactor being designed for producing weapons plutonium and proved the feasibility of creating nuclear weapons with a plutonium charge.

NOTES

1 On March 15, 1946, Sovnarkom was transformed to the USSR Council of Ministers.
2 M.G. Pervukhin and V.A. Malyshev were Deputy Chairmen of the Council of Ministers and the Ministers of the Chemical and Machine Building Ministries, respectively. Malyshev headed the State Commission responsible for the first thermonuclear weapon test carried out on August 12, 1953. He died of the acute leucosis on February 29, 1957.
3 Spacing between the holes drilled in the graphite bricks for inserting uranium slugs.

CHAPTER 4

First Production Reactor A

The first steps in creating an atomic industry were made prior to the organization of the Special Committee and PGU and the construction of the Plutonium Center in the South Ural. The State Defense Committee's decree of May 15, 1945, approved the increase of the production capacity of the NKVD Department No. 9 for a period of 1945–1946. The same decree prescribed to increase the production of uranium ore at Center No. 6 and enhance the capacity of the plants engaged in processing uranium ore and uranium concentrates. It was planned to complete the building and equipping of the NII-9 facility, located in Moscow, by January 1, 1946. This institute was assigned a mission of extracting uranium from ore and manufacturing articles for the experimental reactor of Laboratory No. 2 and for an industrial uranium-graphite reactor designed for the production of plutonium. For secrecy reasons this institute was named Base No. 1 [38].

On August 12, 1945, G.D. Smyth published his monograph on the utilization of atomic energy for military purposes. He wrote: "The cost of the project on the construction of whole cities and unprecedented plants stretching over many miles and an unbelievable amount of experimental research – all this has been concentrated in the atomic bomb as in a focus. No other country in the world was capable to spend such intellectual energy and engineering efforts." Smyth and those in the USA administration who authorized the publication of his monograph were wrong: there was such a country in the postwar period – it was the Soviet Union. This fact was attested by the technologies and equipment designed for Center No. 6 in Central Asia and a number of large enterprises in the Ural region, which were under construction at that time. Since the 1946–1947s the production nuclear reactor was the most important project

**IGOR VASILIEVICH
KURCHATOV,**
Academician from 1943.
**Scientific coordinator of the
Soviet nuclear weapons project.
Director of Laboratory No. 2
from 1943 to 1960 (from 1960 the
laboratory became Kurchatov
Institute for Atomic Energy).**

among the newly built facilities. Prior to its completion, experimental research was carried out using the F-1 reactor at Laboratory No. 2 to substantiate its physical parameters.

The urgency of the measures taken in the initial period of the atomic project can be perceived from the memoirs of V.F. Kalinin, a veteran of the atomic industry, with whom the author had a number of meetings.

Kalinin was recruited to the atomic industry in 1945. He was the first scientific secretary of the Nuclear Reactors Department of the Engineering Technical Council, which later became a department of the PGU Council of Science and Technology. Later Kalinin worked in the Administration of the USSR Minsredmash Ministry. He was one of the pioneers who created the Central Departmental Institute for Information. He translated Smyth's book into Russian using a photographic copy of the manuscript obtained by the Soviet intelligence in the late half of 1945. So, prior to its publication the book was used to draw up a fundamental governmental decree initiating the development of various lines of the atomic program, primarily the building of the first production nuclear reactor near Kyshtym City in the South Ural region.

Dozens of thousands of workers and constructors of various specialties, engineers and researchers belonging to different branches of industry, builders from the NKVD Glavpromstroy Department, and military builders of the Peoples's Commissariat of Defense were engaged in the projects of a dawning atomic industry. Suffice it to say that about 45 thousand workers labored in the 1947–1948s at the construction site of the Plutonium Center. Later, after the PGU was established, governmental decrees dealing with emergency situations in the atomic program were prepared mostly by the officers of the PGU and a Special Committee.

The following three designs of the reactor based on the use of natural uranium were examined at Laboratory No. 2: a heavy-water, a gas-graphite, and a water-graphite reactor. By the middle of 1945, preference was given to a uranium-graphite reactor. The same concept was proposed for a 100-megawatt water-cooled production reactor. The difficulty was to decide between discharging the large amounts of low-active water passing throughout the reactor cooling system into the adjacent Kyzyl-Tyash Lake (area of 19 km^2, water volume of 83 mln m^3) and directing this water through a purification system for a repeated use. The latter program assumed the accumulation of a large amount of radionuclides in the equipment of the purification system. Naturally there was no experience in treating high-active rad-wastes.

On January 12, 1946, the Session of Section 1 of the Engineering Technical Council heard the reports delivered by two experts from the All-Union Thermophysical Institute (VTI), E.N. Romm (water supply projects) and V.G. Prokhorov (water purification projects). The VTI section recommended that a flow-type cooling system be designed for the production reactor[1] and necessary water purification systems be constructed. The same design was also planned for next uranium-graphite production reactors which were later constructed at Center No. 817. That was a forced solution for it was impossible to create a more sophisticated technological design in a short time period under the postwar conditions.

NIKOLAY ANTONOVICH DOLLEZHAL, Academician from 1962. Chief designer of the first Soviet plutonium production reactor, the first reactor destined for atomic submarines, and the first world nuclear power plant.

A team of researchers from Laboratory No. 2 headed by I.V. Kurchatov, who worked out a scientific basis for a future production reactor, included V.V. Goncharov, I.S. Panasyuk, V.I. Merkin, P.I. Shestov, N.S. Bogachev, I.I. Gurevich, V.S. Fursov, B.G. Dubovskiy, N.F. Pravdyuk, M.I. Pevzner, M.S. Kozodaev, S.M. Feinberg, S.A. Skvortsov, and Yu.A. Prokofiev. From 1948 on, most of them, as well as workers from many other institutes and enterprises, worked on a permanent basis at the construction site of a production reactor [11].

N.A. Dollezhal, chief designer of a production reactor, wrote in his notes [39] that in January 1946 he was invited to work for the atomic program by M.G. Pervukhin, who was the People's Commissar of Chemical Industry at that time. B.S. Pozdnyakov, scientific secretary of the NTS Council, and Kurchatov immediately assigned Dollezhal with a task to create a production uranium pile in the shortest possible time. First source data for designing this pile were provided by Laboratory No. 2 proceeding from a horizontal arrangement of fuel channels (FCH) where uranium slugs were to be cooled by water. This design was similar to the military reactors built at Hanford. A reactor intended for the commercial production of plutonium was developed during a cold war period initiated by a speech of W. Churchill in Fulton on March 5, 1946. Very short timescales were set by the Special Committee and PGU Administration: the chief designer was to deliver a technical project and draft drawings of reactor basic components by August 1946. Five teams of designers were immediately set up at the NIIKHIMMASH Institute for working on the project. The teams were headed by P.A. Delens, V.V. Rylin, V.V. Vazinger, B.V. Florinskiy, and M.P. Sergeev. On December 26, 1945, Section 1 of the Engineering Technical Council charged Laboratory No. 2 with devising an alternative vertical design of a reactor. In February 1946 Dollezhal who was devising a reactor design and its technological systems suggested to use a vertical instead of a horizontal arrangement of fuel channels to eliminate problems caused by the deformation of the reactor construction units taking place during its operation: these units ceased to be loaded on heating. This project was approved in March 1946 by a commission consisting of experts in physics, metallurgy, machine-building, chemistry, and of the heads of the atomic project, I.V. Kurchatov, V.A. Malyshev, B.L. Vannikov, M.G. Pervukhin, A.P. Zavenyagin, V.S. Emeliyanov, B.S. Pozdnyakov, and E.P. Slavskiy. A final decision in favor of the vertical design was adopted on July 8, 1946. However, designers from the KB-10 (later a design bureau of the OKB Hydropress) and designers from the GSPI-11 Institute proceeded with the horizontal design of a reactor.

A special sector in hydraulics was set up at the NIIKHIMMASH Institute by a governmental decree. It included designers from other organizations. Moreover, the following institutes and design bureaus were engaged to work on a share basis on the NIIKHIMMASH design project: *Proektstalkonstruktsiya* (N.P. Melnikov, Director), Design Bureau of the Aviation Ministry (A.S. Abramov, Chief), Institute of Aviation Materials (A.V. Akimov, Director), IFKh (Academician A.N. Frumkin, Director), and All-Union Institute of Hydro-Machine Building (department headed by Professor V.V. Mishke). The projects for the reactor building and other facilities of the reactor cooling system were devised by the GSPI-11 Institute. All construction drawings were prepared at the GSPI-11 under the guidance of A.A. Chernyakov, chief engineer of project [39]. This institute was a general contractor in designing all of the plants of the Atomic Center and the city of Chelyabinsk-40.

On April 24, 1946, the NTS Section 1 (M.G. Pervukhin, I.V. Kurchatov, B.S. Pozdnyakov, N.A. Dollezhal, B.M. Sholkovich, E.N. Romm, V.F. Kalinin, etc.) attended by the PGU managers B.L. Vannikov, E.P. Slavskiy, and A.N. Komarovskiy, and also by P.T. Bystrov, director of Center No. 817, approved the general design of Center No. 817 proposed by the GSPI-11 Institute and Laboratory No. 2. It showed the location of the reactor, flow cooling systems, water-treatment and chemical purification facilities, and a site for a settlement. A.A. Chernyakov, chief engineer of the project, was charged with designing the reactor emergency cooling system including the construction of a permanently operating 2-megawatt thermal electric power plant (TETs) and emergency water tanks with a total capacity of 500 m³.

The general scientific guidance and supervision of this work was accomplished by Kurchatov. Dollezhal wrote [39]: "Kurchatov visited the institute every 3 or 4 days. He was usually accompanied by some of the PGU managers or collaborating directors and always by Merkin whom he called his chief technologist."

The NIIKHIMMASH Institute had its own experimental plant and was able to promptly manufacture all necessary testing facilities to verify the operation of various components and units. The efforts of large teams of designers, metal scientists, machine builders, metallurgists, instrument makers, and specialists in corrosion and water-chemical cooling of uranium slugs resulted in a reliable design of production nuclear reactor equipped with all necessary systems for checking and control of a nuclear chain reaction.

Despite the objections posed by the Heads of Section 1, a proposal made by Merkin was approved. He suggested equipping each fuel channel comprising water-cooled uranium slugs with four control systems to monitor a declining or rising water flow rate (SRV and PRV signals), cooling water temperature at the output of the channel, and humidity in the pipe–graphite annulus. In the upper section of each channel a special valve was installed through which the working water (delivered from pumps at the pressure of 8 atm) flowed to cool the uranium slugs in the operating reactor. This valve had two positions, and was able to maintain a minimum water flow through a channel in case of a pressure drop. In the case of an emergency failure of the pumps the reactor was shut down and the no-load water reserve kept in the emergency tanks was delivered through the open valves to the channels to remove residual thermal power. No intermediate position of the valve was allowed for it might result in a sharp drop of water flow through an operating channel and cause the destruction of uranium slugs. Usually operators measured the temperature of water at the channel outlets manually directly using a "peg board" of the control panel (4 operators per shift were engaged in this procedure; those operators, usually women, were called "hot girls" for fun). The air humidity in the pipe-graphite annulus was measured manually at the humidity control panel. Variation of these parameters allowed operators to control the behavior of uranium slugs, their suspension, or the deterioration of the channel tightness. In the latter case prompt measures were taken to withdraw uranium from the channels. In case uranium slugs were jammed (suspended), the reactor was shut down. With the number of fuel channels being above a thousand, one can imagine problems that might occur during the reactor operation, particularly the ones that might jeopardize its operation. To avoid this, in addition to above systems, the SUZ control and safety system was designed under the guidance of A.S. Abramov. It allowed designers to adjust the distribution of power production over the range of the reactor core and ensured the shut-down of a chain reaction in case of emergency PRV- or SRV-signals from the channels or the absence of water in the core.

The vital systems without which the reactor could not operate included a system providing the loading and unloading of uranium slugs and an emergency cooling system initiated in the case of a power supply failure or a water pipeline collapse.

It should be noted that a system for unloading uranium slugs, devised under the guidance of N.A. Dollezhal, was tested at the institute using only one channel. In the course of the mass production of unloading devices, serious defects were found in their design: the unloading operation resulted in the jamming of uranium slugs which got stuck in the channels. This unloading system was rejected. The Gorky Machine Building Plant (GMZ) was urgently charged with a prompt design and manufacture of a new unloading system (Yu.N. Koshkin, chief designer). The new cartridge-type system turned out to be operable and was later applied in all reactors of the same type. The replacement of the unloading system delayed the reactor launching for a few months.

The emergency power supply of the reactor was to be provided by a specially built heat and electric power coal plant with an initial capacity of 2 megawatts. I.P. Lazarev, a veteran of the atomic industry, recalls that this permanently operating plant provided emergency cooling of reactor A, and also other reactors, each time the Uralenergo power supply system failed. Upon increasing the emergency power plant capacity to 10 megawatts, its energy was transmitted to the general power supply line during the normal power supply of the nuclear reactors. V.I. Surkov was the first chief of the power lines and subplants department.

The irradiated uranium slugs were removed from the reactor and kept in specially designed water basins to allow the decay of short-lived radionuclides. After the activity of the slugs dropped they were transported in special container carriages to a radiochemical plant.

A few thousand cubic meters of highly purified cooling water were permanently being pumped through the reactor to provide its proper operation. For this purpose, pumping stations were built, and special facilities were provided for the treatment and chemical purification of water delivered from the Kyshtym Lake located about 1 km from the reactor. All crucial problems of the reactor creation were discussed by Section 1 or at the sessions of NTS Council. The draft drawings of the production nuclear reactor and basic documents produced by the design institutes were coordinated by Kurchatov, approved by the PGU Authorities, and were subject to the urgent implementation by the manufacturing plants. In particular cases the PGU together with the researchers prepared governmental decrees charging the industry with supplying materials necessary for the atomic program.

As mentioned above, to load the production reactor one needed about 150 tons of very pure uranium slugs and no less than 1,200 tons of graphite with negligible impurities of neutron absorbers (boron, etc.).

Within a few months the uranium slugs were to be regularly reloaded after they accumulated certain amounts of plutonium. The graphite components (graphite bricks of $200 \times 200 \times 600$ mm) used for the reactor core and the side and face reflectors should be operative throughout the entire life of the reactor following their assembly and the installation of biological shielding. Thus, one could improve the technology of producing uranium slugs, whereas poor quality graphite would immediately inhibit a controlled nuclear chain reaction in the pile. The tests conducted at Laboratory No. 2 and at the Moscow Electrode Plant (under the guidance of V.V. Goncharov and N.F. Pravdyuk) attested that the graphite quality was good enough to create an operable reactor.

K.T. Bannikov was appointed a scientific supervisor of the graphite production at the plants. E.P. Slavskiy, Deputy People's Commissar, headed the practical operation of the Moscow Plant of the People's Committee of Non-Ferrous Metals which was charged with manufacturing graphite for nuclear reactors from an electrode mass. E.P. Slavskiy recalled in his memoirs [40]: "We did manage to produce pure graphite. The electrode mass was mixed with chlorine and heated to be red hot. At high temperature the interfering impurities bonded with chlorine, became volatile, and were removed. Thus, we finally began to produce pure graphite." Uranium slugs for production reactors were manufactured at Plant No. 12 in Elektrostal City.

A grade classification was used in the production of graphite. The graphite grade was assessed from a physical index based on its capability to absorb thermal neutrons.

Construction and Start-up

In the middle of 1947 General M.M. Tsarevskiy, who had previous experience of building the Gorky Motor Vehicle Plant, Nizhniy Tagil Metallurgical Center, and other large industrial facilities, was appointed the head of the large-scale construction of buildings for the reactor, other facilities, and the city. This appointment was made by L.P. Beriya who visited the construction site on July 8, 1947.

When choosing a site for the future city of Chelyabinsk-40, special investigations were carried out to determine the wind direction, the dilution of harmful materials released to the atmosphere by the reactor and radiochemical facilities, and the optimum sizes of the outlet pipes of production facilities to eliminate hazardous effects on the city. A special survey team headed by V.A. Saprykin, chief engineer of the construction project, prepared necessary recommendations for a rational location of various production facilities and the future city of Chelyabinsk-40. The high management capabilities of V.A. Sapry-

**BORIS GLEBOVICH
MUZRUKOV.**
Director of the Plutonium Center
in the South Urals, from 1947 to
1953.

kin, a future Academician in architecture, were well known [33]. B.L. Vannikov, the deputy directors of the PGU A.P. Zavenyagin and A.N. Komarovskiy, and the managers of the Center were responsible for the general supervision of the construction operations and logistics.

Taking into consideration the extraordinary significance of building a production reactor, the leaders of the atomic program decided to appoint a new head of the construction project instead of P.T. Bystrov, Director of the Center. E.P. Slavskiy was appointed a new director of the Center on July 10, 1947. At the same time, I.V. Kurchatov was appointed a scientific supervisor of the Center.

By the end of 1947 the main building for the first nuclear production reactor was erected, and a field of installation operations was ready. S.M. Piyankov was appointed director of Project A, V.I. Merkin[2] chief engineer, and N.D. Stepanov his deputy. From the end of 1948 through 1952 N.N. Arkhipov was the director of Object A. I.S. Panasyuk,[3] deputy director of Sector 1 of Laboratory No. 2, was appointed the first scientific supervisor of Project A. The Special Committee appointed N.N. Arkhipov, A.D. Ryzhov, A.I. Zabelin, L.A. Yurovskiy, and D.S. Pinkhasik chiefs of operating shifts at A-reactor. Concurrently they acted as duty chief technologists. They managed all operations in the central hall of the reactor and personally took part in them. The reactor servicing staff vacancies were filled up, and necessary engineering services and laboratories were organized.

In late November 1947 Beriya visited the Center for the second time. He appointed B.G. Muzrukov director of the Center. E.P. Slavskiy was appointed chief engineer and held this post by the end of 1949. A great amount of construction operations was executed at the Center at that time in accordance with the GSPI-11 project. At a meeting chaired by A.P. Zavenyagin the construction managers reported the following amounts of work done during a year and a half, by the spring of 1948 [33]: 190,000 m^3 of earth, mostly rocks, were excavated; 82,000 m^3 of concrete and 6000 m^3 of bricks were laid. Because the production reactor was considered to be a very important facility, M.M. Tsarevskiy and V.A. Saprykin were instructed to transfer builders one by one to the radiochemical plant as the reactor construction was coming to an end. In late 1948 this plant was to receive uranium slugs irradiated at the reactor for extracting plutonium [33].

In January 1948 metal structures and basic equipment started to be installed at the reactor site under the supervision of V.F. Gusev, representative of the chief designer. Work on assembling a reactor core and a reflector of graphite bricks began in March 1948. The air coffer was built inside the partially covered building above the core to prevent cold air as well as dust and dirt from the surrounding area to penetrate into the reactor core. Erection and welding were still under way inside the reactor building [11]. A team of physicists headed by I.S. Panasyuk measured the purity of graphite bricks using special neutron probes as they were laid in the core.

By the end of May, 1948, the basic assembling operations were completed, and the testing of the reactor machinery and control systems began. Dollezhal, chief designer of the reactor, spent nearly 5 months at the construction site. To provide an operative management, Kurchatov and Vannikov lived close to the construction site throughout the entire assembling and launching period. As B.V. Brokhivich recalls, a three-bedroom cottage with a firewood heating system was built for them half a kilometer from the reactor site. Close to that house there were a guardhouse and a canteen for the engineers and technicians. "Their residence was included into a restricted area and was guarded by MVD [41]." A.P. Zavenyagin, M.G. Pervukhin, A.N. Komarovskiy, B.S. Pozdnyakov and other officials of the PGU and Special Committee who were responsible for the timely delivery of the equipment regularly visited the Center.[4] Vannikov, Kurchatov, and managers of the Center regularly held operational meetings at the reactor construction site where they discussed the construction of the facilities and the assembling of the equipment. A.N. Komarovskiy, Deputy Director of PGU, described in detail the construction operations, preparation of sites and communications, excavation of foundation pits in hard rocks down to a depth of 40–50 m [35].

Specific requirements were imposed upon materials for biological protection from the ionizing radiation of the reactor facilities. A weighted concrete with a bulk density of 3.6 ton/m^3 was filled with Krivoy-Rog iron ore (hematite) crushed to a required size. Also hematite concrete doped with a conventional mineral sand was used [35]. In addition to the concrete shielding, the reactor had an external shielding consisting of tanks filled with water 1 m deep. The tanks mounted at the uranium-graphite reactors of the Plutonium Center were called "Leonid." The other structures of the reactor were also given code names. For instance the biological shielding above the core was named "Elena," other structures were named "Olga," "Roman," "Stepan," etc.

V.I. Shevchenko, a veteran of the nuclear center, recalls in his memoirs [42] that they mounted 1,400 tons of metal structures, 3,500 tons of equipment, 230 km of pipelines of various diameters, 165 km of electric power lines, 5,745 valve units, and 3,800 sensors.

In 1947 a Central Laboratory (TsZL) was set up. Its personnel was engaged in the improvement of technological processes at the operating facilities and carried out research work in cooperation with the head research institutes of the atomic industry. The first directors of the laboratory were P.A. Meshcheryakov (from 1947 to January 10, 1951), V.P. Shvedov (till February 14, 1952), and V.I. Shirokov (till June 16, 1955). The Central Laboratory included a Production Control & Automation Service (later transformed to a Control Measuring Instrumentation and Apparatus Laboratory and Special Design Bureau − OKB), headed by Yu.N. Gerulaitis and S.N. Rabotnov and a Radiobiological Department, headed by G.D. Baisogolovyi and V.K. Lemberg. Kurchatov had his offices in the reactor building and Central Laboratory. Adjusting the critical mass of uranium, maintaining the purity of graphite and construction materials, the automated control of processes, the design of devices for removing the slugs, disposal of radioactive wastes, radiation protection of personnel − this is an incomplete list of problems which were solved on site under the guidance and with a personal participation of Kurchatov [11]. B.G. Muzrukov, director of the Center, recalled that it had been a personal direction of Kurchatov to start building laboratory premises where a physical and a radiochemical laboratory were accommodated later. Among those who took part in laboratory experiments were the following engineers-researchers of the Central Laboratory and Facility A: E.A. Doilnitsyn, E.E. Kulish, V.N. Nefedov, G.B. Pomerantsev, Yu.I. Korchemkin, V.I. Klimenkov, G.M. Drabkin, A.G. Lapkin, etc.

In early June 1948, after the fuel channels were arranged in the graphite stack, tests were run of the heat-removal system: pumps, delivery and circulation of water through all channels and water pipelines, including water discharge into lake. Thereafter, 24-hour loading of uranium slugs was started.

The work at all workplaces was strictly kept to the rules. The loading of uranium into the reactor core was personally controlled by Vannikov. His workplace was arranged in the central hall of the reactor. The proper loading of the slugs was checked using a special plumb lowered into a channel. One day, F.E. Loginovskiy, deputy director of a shift, who was responsible for this checking procedure, dropped a plumb and a rope attached to it into a channel. Vannikov took away his identity card and told him that if he failed to raise the rope, he would not get back his identity card and would stay in the zone with the prisoners. With the aid of a special device the rope was pulled out from the channel and the reactor operating capacity was recovered. Strictness was a specific feature of Vannikov, sometimes it was expressed in a humorous manner. Once, a worker of a construction organization, whose name was Abramzon, was punished for a poor report and faulty actions during the mounting of equipment. The Director took away his identity card saying: "You are not Abramzon, you are Abraham in a zone," and sent him for some time to the prisoners' camp located close to the facility. At that time the whole of the construction site was like a camp...

The laborious procedure of loading uranium slugs into a thousand channels was implemented with a special care. Among the members of the first team who participated in loading the slugs were Vannikov, Kurchatov and the managers of the plant. The reliable operation of the SUZ control system, the control of the water flow rate and temperature in the channels and radiation situation in the premises were essential conditions for starting up the reactor.

A team of physicists headed by I. S. Panasyuk used start-up tools during the continuous monitoring of parameters controlling the onset of a chain reaction. In the evening of June 7, 1948, Kurchatov took over the functions of a chief operator at the reactor control panel. In the presence of Vannikov, the managers of the plant, the director of the reactor shift, and the engineers on duty, he started an experiment on the physical start-up of the reactor without using the heat-removal system and with emergency rods withdrawn from the core. At 00.30 a.m. on June 8, 1948, the reactor reached a power output of 10 kilowatts. After that Kurchatov shut down the chain uranium fission reaction.

The next phase was starting up the reactor with water in the channels; it took nearly two days. After turning on the water cooling supply, it became evident that the amount of loaded uranium was insufficient for the operation of the reactor. The chain reaction did not begin even with pulled out rods − neutron absorbers. To start up the reactor with water-filled channels, they had to load additional uranium slugs batch-by-batch, with intermissions. Only when the fifth batch of uranium slugs was loaded, i.e., when a nearly 20-percent surplus over the design load was attained, did the reactor become critical with the last control rod withdrawn by 2/3 of its length. This took place at 8.00 p.m., June 10. The same day the power output of the reactor with water-filled channels was brought up to 1 megawatt, and the reactor was run at that power for 24 hours. On June 17, 1948, a warning record was made by Kurchatov in the shift director daily operation log [42]: "Shift Directors! I warn you that an explosion will occur in case water supply is cut off. Under no circumstances should water supply be cut off. As a last resort, you may cut off the operating water flow. The idle water supply should be provided permanently. You should watch the water level in the emergency tanks and monitor the operation of the pumps."

At 12.45 a.m., June 19, the long-term preparation of the reactor for operation at the rated power output of 100 megawatts was completed [7]. This day is taken to be the beginning of the production period of Center No. 817 located in Chelyabinsk-40. The continuous round-the-clock operation of this installation and the permanent accounting of energy output and, also, of plutonium produced for the first atomic bomb, started on that day. Naturally, the quantitative production data were known at that time only to a few persons. The accuracy of the estimated amount of plutonium accumulated in the reactor could not be high, because at that time the cross-sections of various neutrons interacting with ^{238}U nuclei were under experimental study. Moreover, it was necessary to know neutron cross-sections ^{239}Pu, too.

Instructions for the plutonium accumulation procedure strictly limited a maximum admissible content of ^{240}Pu in ^{239}Pu.

Theoretically, it was possible to calculate an approximate yield of plutonium in the reactor. Within a few years the knowledge of its actual yield was experimentally improved at the radiochemical plant. Taking into account the 2.3-day half-life of ^{239}Np, in the first years the uranium slugs irradiated in reactor A were kept for a period of 10–20 half-lives of ^{239}Np before the delivery to the radiochemical plant, to allow all neptunium to pass to plutonium. During this procedure the total radioactivity accumulated in the uranium slugs decreased many times. At the radiochemical plant, special tanks were provided to keep uranium and plutonium-bearing solutions for additional ^{239}Np decay [43].

Problems in Servicing

Physical processes that take place in the reactor are related to the maintenance of a controlled nuclear chain reaction both when operating at a rated power and after the reactor is shut down. Usually the capacity of a shut-down reactor does not drop to a zero – the chain process is not actually shut down, and the reactor temporarily operates at the 0.5–1% of its rated power. There are numerous reasons for shutting down the reactor. Usually shutdowns are caused by a necessity to eliminate or adjust various departures from the prescribed operational mode. During the initial period of the production reactor operation there were many cases of false actuations of the emergency protection system. Shutdown situations lasting 20–60 minutes reduced the production of plutonium. They were always reported to the administration of the Center and PGU.

Running the power output from the 1-percent or zero level to the rated one is a demanding technological operation for the servicing personnel of the reactor, especially for the shift director and the chief engineer responsible for the reactor control. Where one fails to raise the shutdown reactor up to its rated power output, the chain reaction may cease, and it would take more than 24 hours to start-up the reactor and provide the design power output required for plutonium production. What causes the termination of the nuclear fission chain reaction in a reactor? As a result of uranium and plutonium nuclei fission, various fission products are generated, many of them having large neutron absorption cross-sections. The capture of neutrons by long-lived and stable isotopes is usually called "reactor slagging," the capture of neutrons by short-lived isotopes is known as "reactor poisoning." In fact, poisoning is caused by the neutron capture by only one isotope, ^{135}Xe, which has a neutron capture cross-section of about $3 \times 10^6\ \sigma$. It is thousand times higher than the capture of neutrons by ^{235}U nuclei. It was later found that nearly 95% of ^{135}Xe nuclei are not generated during a fission reaction, but result from the beta-decay of another short-lived isotope, ^{135}I. The generation of ^{135}Xe after the fission and radioactive decay of uranium proceeds as follows:

$$\underset{92}{235}U + \underset{0}{1}n \rightarrow \underset{52}{135}Te \underset{19s}{\xrightarrow{\beta^-}} \underset{53}{135}I \underset{6.6h}{\xrightarrow{\beta^-}} \underset{54}{135}Xe \underset{9.1h}{\xrightarrow{\beta^-}} \underset{55}{135}Cs \underset{2.3 \cdot 10^6 \text{years}}{\xrightarrow{\beta^-}} \underset{56}{135}Ba.$$

In a reactor operating in a steady-state regime and at a rated power the ratio between the isotopes ^{135}I and ^{135}Xe is steady. In the case of a reduced power output or a shutdown, ^{135}I having a very low neutron capture cross-section is intensively converted to ^{135}Xe. As seen in the above scheme, the latter decays over a longer period. Thus, it takes a considerable time to reduce the number of ^{135}Xe nuclei to a value characteristic of the reactor previously operating at a rated power. This time period, known as an iodine pit depth, depends on the initial neutron flux density and increases with its growth. The time during which the reactor stays in the iodine pit, i.e., in the state where $K_{eff} < 1$ is a few half-lives of ^{135}Xe. Therefore, in case the operators fail to eliminate the reactor malfunction within a short-term shutdown period (≤ 1 hour), Xe poisoning will spontaneously shut down a chain nuclear reaction. As a result, the reactor will have a zero power output over a day or even more. According to the rules that were valid at that time, these shutdown situations were to be reported to PGU. Usually special commissions investigated faults due to the servicing personnel or inadequate reactor control. Culprits were punished. A case of reactor poisoning with xenon, when a thermal neutron reactor is shut down for a day or more, is still called "dropping into an iodine pit." At that time the iodine pit was like a sword of Damocles for shift directors as well as for the directors of the installation because the monthly production of plutonium dropped 3–5% even in the case of a single shutdown.

The process of reactor slagging process reduces the multiplying capability of a reactor, but does not materially affect its operation mode. However, this process restricts the duration of using the initial uranium load in the core because it increases the undesirable absorption of neutrons by fission products. The isotopes of samarium and gadolinium have the highest values of the neutron capture cross-section: approximately 74,500 σ for ^{149}Sm and 200,000 σ for ^{157}Gd.

However, the major difficulties during the initial operation period were caused not only by the physical parameters of the reactor. Continuous operation of the reactor required a reliable control and safety system (SUZ) and the trouble-free operation of instruments controlling temperature and water flow rate in each of the channels. Thermal impact on the uranium slugs considerably varied depending on the distribution of power production over the radius and height of the core. Different limits were set up for the actuation of the RWI (reduced water intake) and IWI (increased water intake) sensors of the emergency system in different sections of the reactor assembly. The reactor was shut down in the case of unacceptable dynamics in the intake of water used for cooling uranium slugs. The corrosion and erosion of the aluminium pipes of the fuel channels and the uranium slug jackets led to another hazard, radioactive contamination of water. The penetration of water into the graphite stack through the eroded pipes demanded the replacement of the channels and the reloading of uranium slugs. Moisture in the graphite changed its physical properties, and in the case of extensive wetting, a nuclear chain reaction might stop. In that case the graphite had to be dried. This procedure required much time because air could be blown only through a small gap between the graphite and the fuel channel (the diameter of holes in the graphite bricks was 44 mm, and that of the fuel channels 43 mm). The reactor during this period was not operating.

All of the unexpected problems were met with during the first year. Besides, there were radiation accidents which were owed their appearance to so-called "pancakes" where corroded or eroded uranium slugs were caked with graphite. One such accident took place during the first day when the reactor was run at a rated power. On June 19, 1948, V.I. Shevchenko, head of the laboratory, discovered an elevated radioactivity of the air (about 300 doses) in the moisture control area. It was found that in cell No. 17–20 the water flow cooling the uranium slugs was insufficient because of a partially open idle flow gate in the fuel channel. The reactor was shut down. When cleaning the cell, it was found that the eroded uranium slugs were partially welded with graphite. Cleaning lasted till June 30. However, during these eleven days the Special Committee and the PGU managers demanded production of plutonium to continue. Details of the elimination of that accident were described in [42], [44]. On July 25, during a shift directed by N.N. Arkhipov (deputy director N.A. Semenov) another "pancake" was produced in cell No. 28–18 after an RWI signal was registered. They should have shut down the reactor and suspended the production of plutonium. However, a decision was made to eliminate the "pancake" in the operating reactor despite the risk of contamination the premises and irradiation of the shift personnel. To cool the cutting tools and reduce the discharge of aerosols and uranium-graphite dust into the central hall, they pumped water into cell No. 28–18. As a result, the graphite stack was wetted, and the fuel channels were subject to corrosion.

Another problem that was encountered for the first time was the swelling of uranium and graphite under the action of neutrons. These effects were studied under the personal guidance of I.V. Kurchatov, R.S. Ambartsumyan, A.A. Bochvar, S.T. Konobeebskiy and other scientists by the researchers of the Central Laboratory: A.G. Lanina, I.T. Berezuk, V.I. Klimenko, and others.

Another problem was a limited delivery of natural uranium: the uranium mining industry was still under development. It is well known that the uranium used for the first loading of the production reactor was brought from Germany after the war. Yu.B. Khariton recalled [29]: "In 1945 a commission headed by Zavenyagin was sent to Germany. Together with Kikoin we were looking for uranium throughout Germany. At last we managed to find 100 tons at the border with the American zone. This amount allowed us to save a year in creating the first production reactor." Yet, the amount of that uranium was insufficient for loading both the F-1 and the production reactor. So, already in 1948 partial use was made of the domestic uranium mined in Central Asia.

The surface of aluminum pipes used in the first loading of the reactor fuel channels was not anodized. Subsequent to the penetration of water into the graphite stack, an intense corrosion process started at the gra-phite–water–aluminum contact. As a result, at the end of 1948, numerous leaks in the aluminum pipes caused wetting of the graphite stack. The reactor could not be operated. On January 20, 1949, the reactor was shut down for an overhaul [42]. The problem was to replace corroded fuel channels and preserve all valuable uranium slugs. One could try to unload uranium slugs through a specially designed unloading system. However, the pulling of slugs downward through a technological pathway (channel–unloading pit–excavator bucket[5]–coo-ling basin) might produce mechanical damage to the jackets of the slugs making their reloading into the reactor impossible. No reserve uranium was available at that time. One had to make use of the partially or highly irradiated uranium slugs. Zavenyagin suggested withdrawal of the corroded reactor pipes, leaving the uranium slugs inside the graphite channels, and inserting new anodized pipes. However, this attempt failed because the alignment of the uranium column was disturbed: the uranium slugs were displaced toward the wall of the graphite bricks during the withdrawal of the leaking pipes because the latter had interior ribs to ensure centering of the uranium slugs. Engineers from the chief mechanic service devised special appliances which allowed one to withdraw uranium slugs out of the pipes upward to the central reactor hall with the aid of special suction devices. But the executors of this operation could not get away without being overirradiated. The two alternatives were: either to shut down the reactor for a long period of time, which was assessed by Khariton as long as a year, or save the uranium load and reduce the loss in plutonium production. The PGU managers and scientific director preferred the second alternative. The uranium slugs were withdrawn with the aid of the suction devices via the top of the reactor by recruiting all of the male personnel to do this "dirty" operation. The uranium slugs withdrawn were to be inserted into new pipes made of aluminum alloy and covered with a protective anodized coat.

Upon withdrawal of uranium slugs from the reactor, the water-flooded (wetted) graphite stack had to be dried prior to assembling new fuel channels and loading the uranium slugs. Overall 39,000 uranium slugs were pulled out [42]. E.P. Slavskiy, who was the chief engineer of the Center at that time, described this emergency unloading of the reactor [40]. He emphasized that the penetration of water into the pipe–graphite gap was also caused by an inadequate design of moisture alarm system [40]: "To change this system, one had to unload the entire reactor... That was a shocking action." Kurchatov received a large dose of radiation in the course of this operation because the irradiated uranium slugs were stacked near his desk in the central hall of the reactor where he was inspecting them. Slavskiy recalls [40]: "He had no time to inspect all of the slugs. If he had, we would have lost him as early as that." Possible causes of this accident were investigated by Beriya, Head of the Special Committee. Kurchatov, Vannikov, Khrunichev[6] and other managers responsible for the operation of the reactor and its supply of cooling pipes and uranium slugs were subject to interrogation. It was very hard to clear oneself, and penalties could be unpredictable. On his return to the Center, Kurchatov wrote in his report to Beriya [44]: "In addition to our previous message we inform you that by the 1st of February, the dehydration of the facility was accomplished and a uniform distribution of temperature over its cross-section was achieved. Condensate ceased to form... On March 26, 1949, the reactor was started to be brought to design power."

Another series of troubles resulting in the overirradiation of the personnel and in the reactor shutdown occurred during the elimination of "pancakes" and the cleaning of the graphite cells from the dust produced during the breakdown of the sticking fuel channels. Sticking took place for various reasons. Where the cladding of the slugs was breached, uranium corrosion products plugged the slug–pipe wall annulus and inhibited water flow through the fuel channel. In most cases the RWI alarm system allowed time to prevent the slugs from severe jamming in the pipes, using a special chisel for pushing the column of slugs (located above the stuck one) down to the unloading vault. Sometimes the pipes broke, and the uranium slugs were left in the stack without cooling. The reactor was not operated until the uranium was removed from the graphite cell with the aid of a special device. Technologically the manual removal of slugs suspended in fuel channels with the aid of a chisel was a delicate procedure.

When a fuel channel breakdown took place, the reactor usually "fell" into an iodine pit, and its idle period was as long as tens of hours. During a short-term shutdown period it was impossible to take out uranium slugs jammed in the graphite even using a collet designed by Kruglikov, which was capable of removing a few slugs simultaneously [41]. Later another effect responsible for jamming was found − swelling of the uranium core on exposure to neutrons. Sometimes too many slugs got jammed, and the unloading operation had to be done ahead of schedule. These troubles posed serious problems for the researchers. They had to devise a new type of uranium core, find a new aluminum alloy for slug cladding and fuel channels, and improve technological processes used at the production plants. A commission (consisting of A.P. Aleksandrov, R.S. Ambartsumyan, V.V. Goncharov, V.I. Merkin, etc.) headed by Kurchatov charged the VIAM and NII-9 institutes with a mission to study the causes of uranium swelling and find means of reducing this effect. Following the changes in the thermal treatment of uranium rods proposed by the researchers of these institutes (A.M. Glukhov, A.A. Bochvar, G.Ya. Sergeev,

V.V. Titov, etc.), Plant No. 12 started to produce more resistant uranium slugs in early 1949.

There were many departures from the normal technological operation of the reactor. Examples were the jamming of the vessel with irradiated slugs in the unloading vault and the dropping of various objects into the fuel channels. For instance, during one of the shifts (headed by D.S. Pinkhasik), after the jammed slugs were forced down, a combine chisel, a metal rod more than 25 m long and 32 mm across, was accidentally dropped into the fuel channel. To eliminate this fault many unforeseen operations had to be carried out under severe radiation conditions. Prior to the delivery of uranium slugs to the radiochemical plant they were subject to sorting: uranium slugs were separated from dummies that were loaded into the fuel channels below the reactor core. There were about 40% of dummies in each vessel being unloaded from the unloading shaft. Sorting was done in water with the aid of special devices. The uranium slugs were transported for radiochemical treatment by rail in special tanks. The "avial" dummy slugs were transported to specially allocated disposal areas. Sometimes emergency situations occurred in course of these operations.

In the remediation of certain accidents the personnel received unacceptable radiation doses. Of particular danger was the jamming of uranium slugs in the vessels taken out of the discharge pits. Sometimes the impacts of these operations were dramatic. For instance, P.S. Pronin, a welder, died very soon, the fact known to many veterans, operators of the production reactor (F.Ya. Ovchinnikov, N.I. Kozlov, L.A. Alekhin, and others). As seen from Table 3, more than 30% of the operators of the production reactor received radiation doses of 100–400 rem and above in 1949.

Table 3 Radiation doses received by the personnel at the production reactor during the first years of its operation [44], %.

Year	Average dose, rem				Average dose for the facility, rem/year
	25	25–100	100–400	400	
1948	84.1	11.1	4.8	–	19.6
1949	10.7	57.7	31.1	0.5	93.6
1950	52.2	47.2	0.6	–	30.7
1951	74.9	25.1	–	–	18.1
1952	83.9	16.1	–	–	14.9
1953	79.3	18.4	2.3	–	19.6
1954	97.0	3.0	–	–	8.9
1955	95.5	4.5	–	–	9.5
1956	98.7*	0.6	0.7	–	5.1
1957	100.0*	–	–	–	4.2
1958	100.0*	–	–	–	4.4
1959	100.0*	–	–	–	3.3

* Since 1956 more than 90% of workers received less than 10 rem/year.

Operations caused by graphite swelling and uranium slug jamming demanded the gaging of graphite cells and even their reaming using specially designed rods and cutters. These operations were conducted in the production reactor by personnel working in shifts and by workers of the Central Laboratory V.I. Klimenkov, A.I. Malov, Yu.K. Shurupov, and others. It was necessary to keep records of operations conducted in each graphite cell. So a system of case histories of graphite cells and fuel channels was introduced. This work was done by a special group keeping records of the reactor operation. Among those who worked in that group over a few years were G.B. Pomerantsev, a future associated member of the Academy of Sciences of Kazakhstan and Yu.I. Korchemkin, a talented physicist-theorist.

During the operation of the reactor over a few years a great many flaws were discovered in the control systems of the technological process which was continuously subject to improvement. For instance, water flow in more than 1000 channels was measured by individual flowmeters containing mercury. In the course of replacement and repair operations, mercury often spilled and contaminated the premises. New, mercury-free flowmeters were designed, manufactured, and mounted instead.

In the early years the graphite stack of the reactor was cooled with air blowers. Because of a high graphite combustibility the temperature of the graphite stack could not be higher than 330°C. Moreover, at a design power of 100 megawatts the maximum temperature inside the graphite stack was limited to 220°C. This restricted the actual power of the reactor and, therefore, its capacity in production of weapons plutonium. Construction of a nitrogen plant for using nitrogen instead of air increased the admissible temperature of graphite and made the reactor power a few times higher. As a result, the flow rate of water and its temperature at the outlet of the channels were increased too. Unfortunately, the actual radiation doses were brought within acceptable limits only 8–10 years later.

Production Reactor as the First Research Base

The onset of the regular production of radioactive isotopes for the national economy and medical applications coincided with the bringing of the production reactor to its rated power. In 1946 the use of radioactive isotopes, especially in medicine, was under the scientific control of the Radiation Laboratory. Decrees issued by the USSR Council of Ministers (July 10, 1948) and Academy of Medical Sciences (September 21, 1948) charged this laboratory with studying the human effects of radiation and devising applications of radioactive sources in medicine and the national economy. This leading research center was headed first by G.M. Frank and later by the well-known scientists: academicians A.V. Lebedinskiy (1954–1962), P.D. Gorizontov (1962–1969), and from 1969 on by L.A. Iliin, Academician of the Academy of Medical Sciences.

The bringing of the production reactor to the design power in 1948 coincided with the establishment of a special preparation laboratory at the Biophysics Institute. This laboratory treated raw materials (irradiated targets), irradiated with neutrons in nuclear reactors or with charged particles in the accelerators, which were already available at RIAN, Laboratory No. 2, LFEI, and KhFEI at that time, and delivered radioactive isotopes to various organizations of the USSR.

The density of the thermal neutron flux was as high as 10^{12}–10^{13} s^{-1}cm^{-2} in the core of the production reactor. Under these conditions specially-designed target slugs could be loaded instead uranium ones into the production channels of the reactor. These targets contained stable isotopes which were converting to radioactive elements within a short period of time under the action of neutrons. During the first years of the reactor operation, on the initiative of Kurchatov, special reactor cells were assigned for producing ^{60}Co, ^{210}Po, ^{32}P, ^{36}Cl, ^{14}C, and some other radionuclides. Naturally, the production reactor could produce short-lived isotopes for the preparation laboratory. These isotopes decayed during transportation. However short-lived ^{131}I isotopes were produced in the uranium slugs by the order of medical organizations. To do this, the irradiation time was reduced a few times and the dissolution of uranium slugs was conducted at the radiochemical plant. The reactor method for producing radioactive isotopes, which had been mastered since the start-up of reactor A, is still used as a basic method for producing radionuclides. For instance, it was reported at the Second Geneve Conference (1958) that 92 of 110 radioactive isotopes were produced in reactors.

It is known that the production rate, the yield of plutonium and other radionuclides depends on neutron flux density. There were no other neutron sources in the USSR at that time except reactor A. So, various physical experiments were started in production reactor in 1949 according to instructions given by Kurchatov. These experiments included the measurement of neutron cross-sections for various isotopes and the renewal of research into nuclear isomerism, a phenomenon discovered before the war under the guidance of Kurchatov. A specially-designed facility, a neutron selector, was installed at the central reactor hall. Researchers of the Center E.A. Doilnikov, E.E. Kulish, G.M. Drabkin, V.N. Nefedov, as well as the workers of the Central Laboratory and Laboratory No. 2, conducted various measurements there. Later, special nuclear reactors were built for producing isotopes and conducting physical research.

An increase in the number of cells loaded with targets to produce isotopes reduces the number of fuel channels loaded with uranium. As a result, the physical parameters of the reactor degrade, the neutron multiplication factor declines, and plutonium yield drops. The higher power intensity of uranium slugs loaded in the fuel channels increases the energy of neutrons and materially changes their interaction with ^{235}U and ^{239}Pu in a particular energy range (close to 0.3 eV). The cross-section of neutron interaction with ^{239}Pu changes by a factor of 5–8 depending on the energy of reactor neutrons. Therefore, any change in the power of a particular channel, as well as of the reactor as a whole, changes the accumulation rate of plutonium, as well as its burnup both due to the fission of ^{239}Pu and its conversion to heavier and more hazardous plutonium isotopes which adversely affect the fission properties of a nuclear explosive. A difference between the total cross-section and fission cross-section of ^{239}Pu ($\sigma_t - \sigma_f$) controls the accumulation rate of unwanted ^{240}Pu isotope in the reactor. In fact one cannot avoid the production of ^{240}Pu, so the irradiation of ^{238}U in the production reactor was limited to short time periods (a few months).

The energy released in the reactor is mostly produced by the fission of ^{235}U and ^{239}Pu. The bulk of this energy is removed by the coolant, and a few percent of the fission energy are converted to the internal energy of long-lived fission products delivered as components of irradiated uranium slugs to a radiochemical plant.

Apart from the fission of ^{235}U under the action of thermal neutrons, nearly 16-percent of this isotope converts to a long-lived isotope ^{236}U with a half-life of about 2.5×10^7 years by the neutron capture reaction ^{235}U + n → ^{236}U. The ^{236}U isotope generated in the reactor is a good raw material for producing a long-lived isotope of neptunium, ^{237}Np.

In contrast to short-lived isotope ^{239}Np which converts to ^{239}Pu nearly completely, when uranium slugs are kept in cooling basins prior to a delivery to the radiochemical plant, the long-lived isotope ^{237}Np ($T_{1/2} \approx 2 \times 10^6$ years) is always present in irradiated uranium. So one has to separate ^{239}Pu from this isotope as may be required by specifications.

It is known that to produce 1 W · s of energy one should provide a fission of 3.1×10^{10} nuclei. The energy of 1 mW · day is produced by a fission of 1.1 g of ^{235}U. The amount of uranium consumed through a neutron capture reaction is higher (1.31 g). The quantity of radioactive isotopes generated in the reactor is defined by the simple formula [45]

$$M = 1.31 \frac{A}{235} X W t, \text{ g,}$$

where A is the isotope atomic number; X is the isotope specific yield; W is the reactor power, megawatts; t is the effective operating time of the reactor at power W, days. One can estimate the amount of ^{239}Pu produced in the reactor using the formula

$$M = 1.33 X_{Pu} W t, \text{ g,}$$

where X_{Pu} is the plutonium coefficient defined as the ratio of the amount of the produced plutonium to that of the burnt-up ^{235}U. In thermal neutron reactors where natural uranium is used the value of X_{Pu} is usually equal to 0.7–0.8. With

X_{Pu} taken to be 0.75, a reactor produces 100 g of plutonium per day at an operating power of 100 megawatts. Even 3–4 months of continuous operation at this power and a uranium load of 150 tons yields a very small plutonium concentration (average 60–80 g/ton). The concentration of plutonium is higher in the middle of the reactor and substantially lower in its peripheral sections. The energy of neutrons increases, and plutonium coefficient X_{Pu} varies with the reactor power. The yield of other radionuclides can be found measuring the neutron flux and neutron capture cross-section of a particular irradiated material in each target-loaded cell.

A special biological channel was designed to conduct radiobiological experiments in the production reactor. This channel was placed at the periphery of the reactor, at the boundary with the reactor reflector. Its diameter was large enough to accommodate experimental animals.[7] Radiobiologists studied the ultimate parameters of animals' viability depending on the doses of gamma-ray and neutron radiation and, also, the temperature factor and time animals stayed in the channel.

Radiation Exposure of the Personnel

Simultaneously with the start-up of reactor A in August 1948, the PGU administration and the Public Health Ministry issued General Sanitary Standards and Regulations for protecting the health of the personnel working at the facilities of the Plutonium Center. An admissible daily radiation dose during a 6-hour shift was set up to be 0.1 rem, i.e., no more than 30 rem per year. In the case of accidents these regulations admitted a 25-rem exposure for a time of ≥ 15 min [46]. After this exposure a worker was to be subject to a medical examination, get a vacation, or be transferred to another radiation-free job. To provide a timely medical examination of the personnel, a medical department (MSO-71) and local medical posts at all facilities were set up in accordance with a governmental decree. The first chief of the MSO-71 department was P.I. Moiseitsev who earlier worked as chief of the medical department at the Elektrostal Plant No. 12. Later, a special Hospital No. 2 was set up under the MSO-71 department in Chelyabinsk-40 to provide prophylactic examinations and necessary treatments.

The exposure rate of gamma-ray radiation was monitored using a dose metering equipment (area monitor). The individual control of the personnel was implemented with the use of film badges which registered exposure ranging between 0.05 and 3 rem with an accuracy of about 30%. Films were developed after each shift. Having a film badge in his pocket each operator knew a total dose received by his body. At facility A special radiation control service was set up. It was headed by I.M. Roseman who previously worked at Laboratory No. 2 dealing with problems of radiation rate control. At the Public Health Ministry a special system for sanitary and medical supervision was set up. As mentioned above, practically all of the technological operations implemented at the production reactor involve high radiation exposure. As follows from the latest publications (see Table 3), after the initial period of 1948 when only 4.8% of workers received radiation doses above 100 rem per year, in 1949 the radiation situation was highly aggravated. Elimination of emergency situations, primarily caused by the impermissible withdrawal of uranium slugs and the breakage of technological pipes in slug jamming situations, resulted in the increased exposure of personnel averaging 93.6 rem per year. Only after 8 years did the radiation situation become stable: 5% of workers were found to be exposed in excess of the annual average dose. In 1952 a new standard was introduced; it limited the radiation dose per shift to 0.05 rem or 15 rem per year. In emergency situations the same limit of the 25-rem dose for 15 min was permitted. Operators and workers engaged in emergency recovery or in equipment repair were permitted an annual average dose of up to 100 rem. In 1945–1956 a routine was set up to transfer workers, based on the radiation control results, to radiation-free jobs for up to 6 months in case of a total exposure above 45 rem per year, or above 75 rem over last two years. It was only in 1970 that rules were introduced to limit the annual exposure to 5 rem per year. Data characterizing the external gamma-ray irradiation of the personnel over the first 10 years of the A-reactor operation are presented in Table 4.

The highest radiation exposure was found among the workers of the machinery and power supply services, and also among the operators of the reactor central hall.

According to the conclusions of N.A. Koshurnikova, Doctor of Medical Sciences, a leading Russian expert in radiation hygiene, the growth of mortality among the groups exposed to high doses (100 rem over 10 years and 25 rem over a year) can be attributed to the radiation effect. The mortality among individuals exposed to lower radiation doses does not differ from the oncological mortality among the adult population numbering 200 cases per 100,000 individuals a year, i.e., about 6% over 30 years (Table 5).

Table 4   Average total dose of external gamma-ray exposure among various professional groups at the reactor for period 1948–1958 [46].

Professional group	Gamma-ray radiation dose, rem
Radiation control service	107.9
Control and measuring instrumentation service	128.6
Machinery and power service	207.5
Personnel of the reactor central hall	203.8

Table 5 Oncological mortality of personnel servicing the production reactor in 1948–1958 [47].

Index	Total dose, rem/10 year (till 1958)		Maximum dose, rem/year	
	<100	>100	<25	>25
Oncological mortality rate	5.7±0.6	9.4±1.2	5.9±0.6	8,7±1.1

Production reactor A building.

This index was above 6% only among individuals who received high doses (during a year and for 10 years). Among workers servicing the reactor, the occurrence of a chronic radiation syndrome was 5.8% [47].

V.N. Doshchenko [48] described his more than 4-year experience of treating chronic radiation sickness in the closed city of Chelyabinsk-40[8] (MSO-71) and at Medical Center 1 of the Biophysics Institute. One of his conclusions was: "It is absolutely inappropriate to consider radiation hazard without comparing it with the other seven risk factors which are known to be far ahead of radiation in their hazardous effects on human health as evidenced by practice." Based on his statistical data on chronic radiation sickness among 1,355 employees servicing the reactors, some exposed to radiation doses exceeding those presented in Table 4 within 2–4 years, Doshchenko emphasized that mortality had not exceeded 0.2% for the entire 40-year observation period.

The military and political goal to produce fission materials for nuclear weapons (weapons plutonium) in the shortest possible time was achieved by the atomic industry. Yet, the very short timescales and limited reserves of raw materials (natural uranium) did not allow technologies to be devised in detail. Everything was for the first time. The country was just getting out of the war, and the industrial potential was obviously insufficient. All this resulted in the excess exposure of personnel to radiation.

NOTES

1   Subsequent experience demonstrated that this normal operation of the reactor would not contaminate the environment. But after a few emergencies (destruction of uranium blocks) the Kyzyl–Tyash Lake was contaminated with radionuclides above the admissible sanitary norm and was used as a closed water basin.

2   From 1949 to 1954 N.A. Semenov, G.V. Kruglikov, F.E. Loginovskiy, and F.E. Ovchinnikov worked as chief engineers of Project A; later they headed other reactors of the center.

3   Later, V.S. Fursov was appointed scientific supervisor of all uranium-graphite production reactors and B.G. Dubovskiy – supervisor of Project A.

4   V.V. Chernykh, Deputy Minister of Internal Affairs of the USSR, coordinated the work of builders belonging to various departments. Within two years he permanently lived in Chelyabinsk-40 with his family.

5   A vessel used for transporting uranium slugs withdrawn from the reactor to a water basin.

6   Minister of the Aviation Industry.

7   There was a vivarium at the Radiobiological Department of the Central Laboratory. Later, Branch 1 of the Biophysics Institute, Public Health Ministry, was set up on the basis of this vivarium.

8   Now known as Chelyabinsk-65 or Ozersk; 81,600 residents as of January 1, 1990.

CHAPTER 5

First Radiochemical Plant B

Production of fissile materials was the main and most complicated problem in creating nuclear weapons. Plutonium produced in a nuclear reactor must be separated from uranium and high-level fission products. A radiochemical plant (Building 101), for security reasons named Plant B, was constructed to extract plutonium.

Irradiated uranium slugs canned in aluminum jackets were delivered from reactor A to the Plant B. Here they were dissolved, and the jacket material and high-level fission products were removed by chemical methods. At the next phases, plutonium was extracted from a huge amount of uranium and transported to a chemical metallurgical Plant V where a crude metal was produced and then refined metallic plutonium and plutonium components for an atomic bomb were fabricated.

The remaining or reclaimed uranium was thoroughly cleaned from the traces of plutonium and fission products. This uranium had virtually the same isotopic composition as the natural one. At that time it was intended to use the reclaimed uranium for the production of another kind of a nuclear explosive − highly enriched ^{235}U. The first gas-diffusion plant (Center No. 813) designed for ^{235}U enrichment was constructed in the Verkh-Neivinskiy settlement.

The concentrated plutonium solution (separated from uranium and fission products at Plant B) was cleaned again to minimize gamma and beta radiation before delivering it to metallurgists. All radiochemical processes used at Plant B were developed at the RIAN Institute under the guidance of Academician V.G. Khlopin. Another more complex technology was developed at the RIAN: the distribution of matter in liquid-solid, solution-solution, and liquid-gas phases was defined more exactly. The possibility of a complete separation of plutonium from uranium and fission products was demonstrated [49].

**VITALIY GRIGORIEVICH
KHLOPIN,**
Academician from 1939.
Director of the Radium Institute
from 1939 to 1950. Pioneer of the
Soviet radiochemistry and radium
industry.

Khlopin is known as one of the creators of Russian radiochemistry. As early as 1915 he started to do research in radiochemistry at the Radiological Laboratory which was founded by V.I. Vernadskiy at the Petersburg Academy of Sciences. Z.V. Ershova [50], a veteran of atomic industry and an outstanding Russian scientist in the field of radioactive materials, recalls that the experimental treatment of radioactive ore was carried out under the guidance of Khlopin, and the first Russian radium compounds were produced on December 1, 1921. Together with I.Ya. Bashilov, Khlopin organized the production of radium from domestic ore materials. In the pre-war period Khlopin headed the Uranium Commission which prepared, as early as that time, a preliminary program for obtaining enriched ^{235}U and utilizing nuclear fission energy. By his obsession and persistence Khlopin resembled the discoverers of radioactive elements. He used the experience gained by Marie and Pierre Curie in the extraction of radium from radioactive ore. Many of his publications, including a review of radium research in Russia published in 1947, were devoted to the production and utilization of radioactive elements.

In 1929–1933 uranium ore processing was initiated at the Moscow Plant of Rare Metals under the scientific guidance of Khlopin and Bashilov. In late 1931 hundreds of tons of uranium ore were processed to produce 200 mg of radium in the form of mixed radium and barium bromide. The certificates of tubes containing radium were attested by checking against the International Standard of the Vienna Radium Institute. During the first years of production they were signed personally by Khlopin. The uranium ore was mined at the Tuya-Muyun Mine. In 1936, after the explored ore was worked out, the production of radium in Moscow was terminated.

The RIAN was, therefore, the only institute in Russia that was able to arrange plutonium extraction from highly radioactive materials and purge uranium of radionuclides on a commercial scale. This organization integrated high-skilled radiochemists, physicists, radiation operators, and other experts in radioactive materials and in 1944 was assigned to develop a radiochemical production technology. Academician Khlopin was appointed a scientific supervisor of this project.

In 1944–1945, after a break for the war period the cyclotron resumed its operation at the RIAN, and the first radiochemical experiments with irradiated uranium were carried out using small amounts of neptunium and plutonium. In 1945–1946 the first research technological report on the methods for processing irradiated uranium was issued. A technology proposed at the RIAN was first tested using a close chemical analog of plutonium — a short-lived ^{239}Np isotope with a half-life approximately equal to 2.3 days. This plutonium analog allowed one to detect a microconcentration of an element in solution using a radiometric technique. This research demonstrated the advanced capabilities of the Russian radiochemistry school created by Khlopin and his students. Among those who held the key positions were B.A. Nikitin, A.P. Ratner, V.I. Grebenshchikova, I.E. Starik, K.A. Petrzhak, V.M. Vdovenko, and Khlopin's companions, B.P. Nikolskiy and A.A. Grinberg.

First portions of plutonium (10^{12} atoms[1]) were produced at Laboratory No. 2 by I.V. Kurchatov as early as 1945, prior to the construction of a nuclear reactor. To this end, targets of uranium hydroxide were irradiated with neutrons produced by a radium-beryllium source. At nearly the same time the first very small quantities of neptunium and plutonium were produced at the RIAN Institute by irradiating targets in the cyclotron. After a more powerful cyclotron was put in operation in 1945 and an F-1 reactor in late 1946, Laboratory No. 2 was able to produce plutonium in considerable amounts [11].

Z.V. Ershova [43, 50] wrote: " ...a research report of the Radium Institute known to all radiochemists as a "blue book" (because of its deep blue binding) was scrutinized by chemists, technologists, and physicists dealing with this problem. This report was used by many chemists and technologists who previously had nothing to do with radiochemistry and radioactive elements. For a long time this report was their handbook." This report and a report presented by Khlopin at a meeting of the Scientific Technical Council (NTS) of the PGU motivated the supervisors of the atomic project to start up the construction of a radiochemical plant in Chelyabinsk–40 as early as summer 1947. Following Khlopin's report, the NTS Council obliged all supervisors (V.S. Emeliyanov, I.V. Kurchatov, V.B. Shevchenko, and others) to speed up the construction and start-up of an F-1 reactor and a subsidiary experimental radiochemical shop. The latter was intended for testing a technology proposed by the RIAN for extracting plutonium from uranium slugs irradiated in an F-1 reactor. Starting from late 1946 this shop (later called Facility No. 5 — one of the NII-9 divisions) was used to perform experiments aimed at testing technologies developed for Plant B which was then in the process of construction. At the same NTS session a report was made by Ya.I. Zilberman, a chief designer of Plant B (GSPI-11). The 1st of March, 1947, was set up as a deadline for Kurchatov to start a pilot radiochemical processing of uranium slugs irradiated in the F-1 reactor and enriched in plutonium. Together with Khlopin, Emeliyanov, and Shevchenko, Kurchatov was given 10 days to draw technical specifications for uranium slugs to be subject to radiochemical processing. According to a proposal made by M.G. Pervukhin and approved by Khlopin, the NTS recommended Plant B to use an acetate–lanthanum–fluoride technology, which was selected from four techniques proposed for radiochemical processing of uranium slugs.

At that time the RIAN team conducted extensive studies of plutonium chemical properties using the small amounts of this element produced at the cyclotron. The cyclotron operation was supervised by M.G. Meshcheryakov.[2] Radiometry studies were headed by K.A. Petrzhak, one of those who discovered a spontaneous fission of uranium in 1940, and by B.S. Dzhelepov and G.N. Gorshkov. Radiochemical techniques were devised under the supervision of B.A. Nikitin, I.E. Starik, A.P. Ratner, and other scientists from the RIAN.

Soon after the publication of the Russian translation of Smyth's book [36], the two-volume monograph *Scientific and Engineering Basics of Nuclear Power Engineering* [52] was published where the main difficulties of solving the atomic problem in the USA were discussed. In the section *Technology for Nuclear Fuel Production*, it was emphasized that in extracting ^{239}Pu aluminum jackets should be removed from uranium slugs by chemical dissolution rather than by mechanical methods. The authors recommended that all technological operations dealing with hazardous fission products be carried out at a considerable distance from populated areas. Another recommendation concerned the construction of special-purpose voluminous storage facilities for radioactive waste. High radiation emitted by fission products was comparable with that produced by many kilograms of radium and hence required the use of remote controls in all chemical operations. Apparently the above publications motivated Khlopin to recommend the following list and sizes of shops at the radiochemical plant under construction:

- chemical preparation shop (12×40 m);
- remotely controlled shop (40×65 m);
- servicing personnel shop (15×80 m);
- plutonium refining shop (15×80 m);
- storage for wasted radioactive solutions ($15,000$ m^3/year);[3]
- wasted solution neutralization station;
- lanthanum reclamation shop;
- hydrofluoric acid reclamation shop;
- warehouses and auxiliary shops.

Because of the complexity of radiochemical technology and the extremely short terms of the construction and launching of Plant B, the Scientific Council of the RIAN recommended the PGU to invite cooperation from other institutes. The proposal of the RIAN was considered on November 1, 1946, at a session of Commission No. 1 established under the Chemical-Metallurgical Department of the NTS and consisting of B.A. Nikitin, A.N. Frumkin, I.I. Chernyaev, I.V. Kurchatov, and delegated A.A. Grinberg and S.Z. Roginskiy. To enhance the coordination and define the responsibility of each research institute at various stages of technology development for Plant B, the commission assigned the responsibilities of the project supervisors in the following way:

- B.A. Nikitin, RIAN, – estimation of the end product (weapons plutonium) output throughout the whole technological process;
- A.A. Grinberg, RIAN, – study of reduction-oxidation processes involved in the extraction of the end product;
- S.Z. Roginskiy, IFKh, – study of radiation effects on the technological operations producing the end product and on the behavior of fission fragments;
- B.V. Kurchatov, Laboratory No. 2, – study of the sorption of end products and fission fragments in the process of obtaining the end product: in precipitates, during filtration, etc.

Many of conventional chemical reactions vary under high ionization conditions:

- the corrosion of construction materials and equipment speeds up;
- explosive peroxides are generated in water solutions;
- organic substances are composed and polymerized;
- radiation energy heats up solutions and hampers temperature measurements during technological processes.

These factors were responsible for serious problems that occurred during the first year of the plant operation (1949).

To test individual technological processes developed at the RIAN, an Experimental Facility No. 5 was built in 1946 in Moscow near the F-1 reactor at the newly established NII-9. It was put into operation in 1947. This facility, combined with a pilot reactor of Laboratory No. 2, was capable of testing the technology, equipment, and control systems of the future radiochemical plant using uranium slugs irradiated in the F-1 reactor. Because of the low concentration of plutonium in uranium, the facility had no biological shield. A group of RIAN researchers was assigned to operate the facility; together with the NII-9 personnel and specialists from other institutes they tested technological processes that were approved by the NTS. A general scientific supervision of this work was provided by B.A. Nikitin, Deputy Director of the RIAN, with the assistance of A.P. Vinogradov, B.P. Nikolskiy, A.P. Ratner, I.E. Starik, Z.V. Ershova, and V.D. Nikolskiy.

An especially important problem for Plant B was the arrangement of sufficient operational ventilation. It was necessary to determine whether it was safe, without a hazardous effect on the environment and population, to discharge into the atmosphere gaseous fission products such as iodine, xenon, and the like, generated during the dissolution of uranium slugs. It was well known that the most dangerous isotope ^{131}I had a half-life of about 8 days. Though a preliminary keeping of irradiated slugs in the water basins of the industrial reactor reduced its concentration by a factor of 15–30, it was insufficient for the full

BORIS ALEKSANDROVICH NIKITIN,
Corresponding Member of the USSR Academy of Sciences from 1943. Khlopin's Deputy, in 1949 coordinated the start-up of the first Soviet radiochemical plutonium-producing plant.

decay of a ^{131}I radionuclide. In late 1946 the NTS Section 4 held a meeting attended by E.P. Slavskiy (PGU), V.G. Khlopin and B.A. Nikitina (RIAN), I.V. Kurchatov and V.I. Merkin (Laboratory No. 2), S.Z. Roginskiy, V.I. Spitsin, and O.M. Todes (IFKh), A.Z. Rotshild (GSPI-11), Z.V. Ershova (NII-9), and leading specialists from NII-26 (Minkhimprom). GSPI-11 and IFKh reports concerning a project for Plant B ventilation system was discussed and approved as a basis. It was assumed that gases produced in the process of the dissolution of uranium slugs would be discharged into the atmosphere through a 130-meter outlet pipe after preliminary dilution with air inside the pipe. To separate radioactive iodine from other gaseous fission products, a sorption technique was proposed by S.Z. Roginskiy (IFKh) and B.A. Vaskovskiy (NII-26). Silica gel impregnated with silver nitrate was suggested to be used as a sorbent. The institutes were commissioned to devise methods for entrapping remaining radionuclides. These and many other problems were solved in 1947–early 1948.

The problem of complete neutralization of large amounts of radioactive solutions produced in the process of plutonium extraction turned out to be insoluble at that time. During two days (25th and 26th of July, 1947) the NTS Section 4 discussed the reports on the problem of liquid radwaste presented by the researchers from the head institutes S.Z. Roginskiy (IFKh) and I.E. Starik (RIAN). Based on the analysis of the state-of-the-art methods in this field, the conclusion was drawn that it was impossible to reduce the concentration of radionuclides in the disposed solutions below 10^{-7} Ci/l. Thus, the disposal of waste into an open drainage system was unavoidable. A request was addressed to the Public Health Ministry of the USSR to inform the GSPII-11, a chief designer of Plant B, about admissible (tolerable) doses in partially radioactive solutions dumped into water basins located in the construction area.

By the summer of 1948, after experiments were conducted at Facility No. 5 and additional studies made at research institutes, the RIAN, NII-9, IFKh, GEOKhI, and IONKh teams developed an updated technology for radiochemical Plant B. Among those who participated in this work were: A.N. Frumkin, P.A. Rebinder, V.I. Spitsin, and S.Z. Roginskiy (IFKh), V.B. Shevchenko,

Z.V. Ershova, and V.D. Nikolskiy (NII-9), B.N. Laskorin and Ya.I. Kogan (NII-26), and other scientists.

M.V. Gladyshev [43] mentioned that at that time there was no adequate technological culture of handling radiochemical products. No due control of the radioactive contamination was put into practice: operators worked without protective clothing and breathing apparatuses. E.P. Slavskiy, the chief engineer of the plant, suggested that training courses should be arranged for Plant B personnel. Nikitin and Ratner developed a training program that included a 70-hour course for engineers and a 59-hour course for technicians. Training took place at the RIAN, the Chemical Department of the Leningrad State University, and at the Moscow NII-9 Institute. Many researchers, who later headed shops and services of the radiochemical plant, worked on probation at Facility No. 5 (NII-9) under the guidance of Z.V. Ershova, head of a laboratory, and V.D. Nikolskiy, Doctor of Chemistry, who were transferred from the State Research Institute of Rare Metals (GIREDMET) in early 1945. Some researchers were trained and worked on probation at the RIAN.

A comprehensive report on the research done at Facility No. 5 was presented by Nikitin on July 14, 1948, at the session of the RIAN NTS Council chaired by Khlopin. The NTS appealed to the PGU administration to approve the project of Plant B. By that time most of the engineering staff were adequately trained, and the construction of Plant B and the installation of equipment were close to completion. Among the future managers of the radiochemical plant, who worked at Facility No. 5, were B.V. Gromov, the first chief engineer of the plant, M.V. Gladyshev and A.I. Pasevskiy – Gromov's deputies, N.S. Chugreev, chief of the department of the final extraction of plutonium concentrate by fluoridation, N.G. Chemarin, chief of the technological laboratory, and Ya.P. Dokuchaev, chief of a group for radiometric control of Plant B processes. The operation of Facility No. 5 was supervised by the administration of NII-9 and supported by researchers from Laboratory No. 2 (Pevzner, Kurchatov, and others), who worked in close cooperation with NII-9 and other institutes.

Tests conducted at Facility No. 5, NII-9, verified the efficiency of plutonium extraction technology developed at the RIAN. The results of the tests were used by the GSPI-11, the institute where the radiochemical plutonium production plant was designed.

Designing and Construction

The Leningrad GSPI-11 Institute started to design Plant B in 1946 using the RIAN source data which continued to be improved until mid-1948.[4] The first version of construction specifications was submitted as a joint project by the RIAN (V.G. Khlopin, B.A. Nikitin, and A.P. Ratner) and the GSPI-11 (Ya.I. Zilberman and N.K. Khomanskiy) as early as the first quarter of 1946.

**ZINAIDA VASILIEVNA
ERSHOVA (1904–1995).
Follower of Marie Curie. Head of
a NII-9 laboratory from 1945.**

The administration of the GSPI-11 (A.I. Gutov and V.V. Smirnov) established a Special Design Bureau No. 2 and appointed L.A. Sytin as a director and A.A. Khonikevich as a chief engineer. The designing was done by A.Z. Rotshild, Ya.I. Zilberman, M.A. Khodos, A.N. Kondratiev, V.A. Khokhlov, M.V. Iolko, E.V. Starobin, L.N. Zhukova, and others. The projects of particular departments of the plant were supervised by V.V. Smirnov, A.A. Chernyakov, A.Z. Rotshild, A.V. Gololobov, V.A. Kurnosov, and others. As the processing technology was improved, the resulting technical specifications were revised by RIAN and Plant B experts. At the early design phase the RIAN participated in all divisions of the project including the technological process, control and measuring instruments, selection of building materials, problems of chemicals regeneration and purity, removal of radioactive products from waste solutions, and personnel protection measures.

The concurrent research lines included the development of techniques for entrapping radioactive aerosols, radioactive gases and volatile radioisotopes and methods for waste detoxication and the extraction of valuable radionuclides from fission products. Subsequent to their irradiation in the reactor the uranium slugs were highly radioactive. The activity of one ton of irradiated uranium slugs was equal to hundreds of thousand gram equivalents of radium. A major portion of this activity was produced by short-lived radionuclides. Besides, the uranium withdrawn from the reactor contained ^{239}Np which decays and converts to ^{239}Pu (basic component of weapons plutonium) with a half-life of about 2.3 days. Therefore, after its withdrawal from the reactor, irradiated uranium should be "cooled" to reduce its radioactivity and increase the abundance of plutonium. Radionuclide ^{131}I (with a half-life of 8 days) is the most dangerous and hard-to-trap product. To decrease the ^{131}I concentration by three orders of magnitude[5] the "cooling" time of irradiated uranium must be no less than 120–140 days, a period during which the overall radioactivity of irradiated uranium decreases by a factor of hundreds.

To reduce the contamination of adjacent areas, the project assumed the construction of an outlet pipe, the highest one in the Ural region (151 m in height[6]), 11 m and 6 m in diameter at the bottom and top, respectively).

Difficulties encountered in the process of plant construction and the installation of equipment were discussed in [43] and [54].

The construction of the plant building and the installation of equipment were supervised by V.A. Saprykin, Chief Engineer of the construction operations. I.A. Anufriev, Deputy Minister of special erection operations, was in charge of the general supervision of installation work.[7] The construction and installation operations were performed by teams from the Construction-Installation Departments (SMU); many of them had participated in the construction of reactor A. Tens of plants and institutes of the USSR were engaged in manufacturing equipment, numerous instruments, remote control systems, and various corrosion-resistant materials. Plant B, the Atomic Center as a whole, and Chelyabinsk-40 City were built by civil and military builders, and, also by numerous prisoners. The overall number of those who participated in the construction of the Atomic Center in 1945–1948 was as great as nearly 45 thousands.

Some of the veterans of the atomic industry (for example, Yu.B. Khariton and Yu.N. Smirnov) believed that the crucial progress in implementing the Uranium Project was achieved only after L.P. Beriya was personally put in charge of the atomic problem: if this large-scale project were still supervised by V.M. Molotov (and M.G. Pervukhin), one could hardly expect progress to be as quick as that [44]. However, it should be noted that Molotov and Pervukhin were in charge of the Uranium Project during the war time (1943–1945) when all resources of the country were consumed by the war. In the late half of 1945, after the USA dropped atomic bombs at Hiroshima and Nagasaki and demonstrated their willingness to dictate to the world their demands, the attitude to the Uranium Project in the USSR changed radically. Undoubtedly the most important role in solving the atomic problem during the postwar period was played by a Special Committee, and Pervukhin was one of its members. It is pertinent to cite a statement made by Kurchatov as early as 1958: "The US aircrafts dropped two atomic bombs at the Japanese cities Hiroshima and Nagasaki at the end of the war, when Germany had capitulated and the military power of Japan collapsed. More than 300 thousand lives were lost in explosions and fire, 200-250 thousand peaceful residents were wounded and exposed to radiation. The US politicians needed those victims to start up an unprecedented atomic blackmail and cold war against the USSR."

During the postwar period the Special Committee concentrated the major resources of the country on solving the atomic weapons problem. Yu.B. Khariton recalled: " ...we got everybody and everything." This situation was impossible during the war, even if the project had been headed personally by Stalin. It took about two years and a half to construct and put into operation two huge and unique production facilities (the first industrial reactor and the first radiochemical plant) at the Plutonium Center and to make ready the launch of Plant B designed for manufacturing the final product. By mid-1948 all auxiliary facilities designed to support the operation of the three major plants were constructed and put into operation: thermal power stations, water-treatment plants, a railroad shop, a locomotive depot, a central plant laboratory (TsZL), mechanical repair shops, etc.

Initial Operation Period

Plant B was put into production in late 1948. B.A. Nikitin, a Corresponding Member of the USSR Academy of Sciences, Deputy Director of the RIAN, was in charge of the start-up team of researchers from all of the institutes that participated in the previous operations at the radiochemical plant. A.P. Vinogradov, a Corresponding Member of the USSR Academy of Sciences, Director of the GEOKhI, was a deputy manager of the start-up team and an assistant of Kurchatov in the analytical chemistry [54]. Another assistant of Kurchatov was Professor A.P. Ratner, Doctor of Chemistry, a coauthor of the above mentioned "blue book" that described the technology to be implemented at Plant B.

The technology implemented at Plant B was based on the acetate-fluoride process. Concurrently, a special-designed facility was installed at Building No. 102 to test a more advanced plutonium extraction process. From the first months of the operation, problems with the corrosion resistance of the equipment arose. G.V. Akimov, a corresponding member of the USSR Academy of Sciences, Director of the IFKh[8] proposed to use nichrome as a construction material for process units. Because the chemicals used were highly corrosive, some of the process tanks, including chemical reactors, were later coated with silver and gold [43].

Irradiated slugs were first charged into the A-201 dissolution unit of Plant B on the 22nd of December, 1948. Yet, the first ready-to-use product was manufactured at the final conversion of the plant in February 1949. Professor B.P. Nikolskiy[9] was put in charge of the process of dissolution of uranium slugs. Later, after A.P. Ratner moved to the RIAN, he was the scientific supervisor of the radiochemical plant.

To clean the extracted plutonium from fission products in order to make their content millions of times lower, the uranium solution was subject to an acetate reprecipitation procedure to separate plutonium from uranium and fission fragments. The resulting plutonium concentrate was treated with fluoride solution in order to remove the remaining impurities by precipitation. Using different reduction capacities of plutonium and uranium, they were first oxidized with potassium bichromate to a hexavalent state in nitric acid and precipitated as salts. The solution still containing macro-admixtures and fission products was discharged, and the residue was dissolved, reduced with bisulfate, and again precipitated with the aid of acetate. Uranium retained its hexavalent state, while plutonium was converted to a tetravalent form and remained in the solution. After filtering, the uranium salts were precipitated on a filter, and plutonium remained in the solution. This procedure was incorporated in the first version of the industrial radiochemical technology [43, 49]. The same principle of separating uranium and plutonium was used in repeated cleaning (in refining) except that precipitation took place in the nitric acid instead of the acetate environment in the presence of fluorine. The solution was then oxidized with bichromate, and hydrofluoric acid was added. This produced a precipitate of rare metal and lanthanum fluorides: lanthanum was added to the solution prior to precipitation. In this oxidized environment plutonium remained in the solution, whereas fission products and lanthanum were removed with the residue. The solution was then reduced by bisulfate, and plutonium was precipitated after the addition of lanthanum.

ALEKSANDR PAVLOVICH VINOGRADOV,
Academician from 1953.
From 1945 — Kurchatov's Deputy for the analytical control of chemical and metallurgical productions plants of the atomic industry. Director of the Institute of Geochemistry and Analytical Chemistry, USSR Academy of Sciences, from 1947 to 1975.

This process technology was complex: the processing tanks varied from hundreds to thousands of liters in size, and hundreds of pipelines, valving components, various devices and level gages were used. The volumes of acids and other chemical agents were many times higher than the amount of uranium loaded into the equipment units. The content of plutonium in uranium was as low as 0.01%, i.e., tens grams per 1 ton of uranium.

The initial operation period revealed numerous difficulties in maintaining the process and many imperfections of the project which were remedied concurrently with the mastering of the technology. Solutions happened to get into the exhaust ventilation, so that plutonium could be found neither in the solution nor in the residue. The large overall area of the surface of the processing units, pipelines and valves was responsible for the loss of plutonium because of its sorption on the walls of the tanks. Unforeseen difficulties emerged in locating plutonium and generally in maintaining the processing plants. The aggressive activity of the reagents caused the corrosion of the equipment and valves, degraded their seals, and complicated the working conditions for the operators. The repair services of the plant, primarily mechanical engineers, instrumentation technicians, specialists in corrosion, and the personnel of the analytical laboratory were operating under emergency conditions and were exposed to unacceptable radiation doses. Suffice it to say that in 1949 the average exposure dose per each worker of the plant was as much as 48 rem/year. In 1950–1951 the technology was improved, the contaminated, partially ruined equipment was replaced by more advanced units, control methods were improved, tools and sensors replaced too. New valves and apparatus were fabricated using low-corrosion material. All this resulted in increased radiation doses: in 1959–1951 an average annual value was around 100 rem [46, 47].

Table 6 lists data on the exposure of the personnel during a period of 1949 to 1962. Only by 1962 the average annual individual exposure dose at the plant was reduced below 10 rem. In 1962 about 50% of the personnel received a yearly radiation dose of no more than 5 rem; only a few persons got 25 rem/year.

Table 6 Radiation doses to personnel at the first radiochemical plant for extraction of plutonium from uranium slugs irradiated in reactor A, in percent of the total number of workers [47].

Year	Radiation dose, rem				Average radiation dose, rem/year
	25	25–100	100–400	400	
1949	26.9	66.2	6.9	–	48.0
1950	21.5	42.0	36.0	0.5	94.0
1951	13.8	41.6	42.8	1.8	113.3
1952	21.8	57.0	21.2	–	66.0
1953	50.7	47.3	2.0	–	30.7
1954	70.8	29.1	0.1	-	20.0
1955	66.5	33.2	0.3	–	21.3
1956	76.9	23.1	–	–	16.2
1957	74.4	25.5	0.1	-	17.5
1958	90.9	9.1	–	–	10.8
1959	96.8	3.2	–	–	14.7
1960	100*	-	–	–	15.2
1961	100*	-	–	–	11.0
1962	100*	-	–	–	7.6

* The radiation dose above 5 rem/year was received in 1960–1962 60, 37, and 50.9 % of the personnel, respectively; 5 rem/year is presently a maximum permitted dose in the atomic industry.

Under demanding conditions the personnel of Plant B, the administration of the Atomic Center and the PGU, and the scientific supervisors of the Atomic Project and radiochemical technology were bound to produce plutonium in early 1949. It was destined for the next technological cycle (Plant V) and to be produced in a quantity sufficient for the fabrication of an atomic bomb (it was exploded at a test site in late August 1949). This quantity was kept secret not only from the workers of the plant and center but even from the heads of the institutes. M.V. Gladyshev [43] who participated in the production of the first portion of plutonium at Plant B recalled: "The first portion of final product was obtained in paste form. I remember how in February 1949 Chugreev and me scraped it off with a spoon from the Nutsch filter located in a special canyon. Though it was difficult to extract plutonium from various impurities, we managed to do this by repeatedly cooking it to an alkaline pulp, dissolution, and washing. The first portion was produced in the presence of the researchers and chiefs of administration in the basement compartment which was for unknown reasons called "canyon." The paste was packed into an ebonite box and delivered to the consumer (Plant V). We did not and were not supposed to know how much plutonium was in that portion. Even later, when I was a chief engineer, the amount of plutonium produced according to a schedule was known only to the Head of the Project. For secrecy only one copy of all documentation was made."

Within a few years the radiation doses were reduced through the use of advanced technology and highly efficient instrument control. This work was headed by E.P. Slavskiy and later by G.V. Mishenkov and done by the engineering personnel of the plant and by researchers from RIAN, NII-9, IFKh, and other institutes. A considerable contribution to the elaboration of the technology was made by the workers of Plant B Central Laboratory (TsZL) N.G. Chemarin, V.I. Zemlyanukhin, V.M. Tarakanov, I.A. Ternovskiy, L.P. Sokhina, and Ya.P. Dokuchaev, to name but a few, and also by the designers of a special design bureau arranged at the plant under the guidance of Yu.N. Gerulyaitis, G.G. Popov, and S.N. Rabotnov. Later the technical design of the plant was subject to considerable changes. Its most complicated department No. 8 was replaced with new equipment installed in department No. 26 where the final product was a nitric acid solution of plutonium nitrate with a known content of impurities. Facilities were provided for the regeneration and purification of radionuclides from some chemical agents to allow their recycling, and measures were taken to reduce the disposal of radionuclides beyond the limits of the plant area.

Recent publications stating that US secrets concerning the creation of atomic weapons were delivered to the USSR [44] attest that the US data did facilitate the development of an atomic bomb in the USSR. However, these secrets were of no use in producing a nuclear explosive – plutonium. All technological processes, materials, and equipment necessary for the radiochemical industry were created in the USSR.

Efforts to Decrease Personnel Exposure

After starting up the F-1 reactor, Facility No. 5, and the first industrial reactor, it became evident that rules were needed to regulate the operations of the personnel under artificial radioactivity conditions. Prior to the start-up of Plant B, in August 1948, the USSR Public Health Ministry (Minzdrav SSSR) and respective PGU departments issued the document *General Public Health Standards and Rules for the Protection of Personnel Engaged at Facilities A and B*. Prior to the start-up of reactor A and radiochemical plant the industry produced dosimeters designed to control the exposure dose rates of gamma radiation. To provide prompt measurements of the daily doses of Plant B personnel, a special laboratory was set up at Plant B, designed to process 2,500 individual badge dosimeters per day [47]. The dose rates were measured using a German-made X-ray film with a sufficient sensitivity. The individual control of the external exposure of the personnel to gamma radiation was accomplished using film dosimeters which measured doses of 0.05 to 3 rem in the energy range of 0.4 to 3 MeV with an accuracy of 30%. Based on the then dominant conception of a tolerable (admissible) dose, the admissible daily dose was chosen to be 0.1 rem per 6 working hours (approximately 30 rem per year). In the case of emergency a one-shot exposure in a zone with a radiation level of no more than 25 rem during ≥ 15 min was allowed, but an exposed worker was to be subject to medical examination and granted a vacation or transferred to a radiation-free job [47]. M.V. Gladyshev [43] recalled, however, that in the first months of the plant operation there was no exposure control, and nobody knew the actual doses received by the workers and engineers of the plant.

The processing equipment was installed at the B-plant in such a way that virtually all of the compartments were subject to radioactive contamination. Because of the corrosion and sealing deterioration of the equipment, as well as for some other reasons, in addition to gamma-active fissile components, products that escaped into the servicing compartments contained considerable amounts of alpha-activity, isotopes of plutonium included. These products contaminated the air inside the servicing areas and were inhaled by the people, thus increasing their unrecorded equivalent exposure doses. After extensive studies, methods were found for protecting man's respiratory tracts against radioactive aerosols. Under the guidance of I.V. Petryanov,[10] individual protective devices, so-called petals, were constructed. They were made of a special fabric called by his name. Their application was mandatory at all production facilities and other places where operations with "dusty" radioactive materials were conducted.

The danger of working under high-radiation conditions was understood by all workers at the Plutonium Center − from its Director to rank-and-file workers at Facilities A, B, and V. However, people believed that the country could not manage to create nuclear weapons without their selfless labor. That was the reason why they risked their safety. Sometimes risks were caused by the necessity to correct their own errors or infractions of the technological regulations, which were severely punished at that time. For some violations people were dismissed or brought to trial.

In 1949 regulations were introduced to set up control of the workers' health. Among other things, they prohibited any hazardous radiation operations without a written permission from the administration and demanded that most dangerous operations be accomplished in the obligatory presence of a radiation supervisor. Despite these measures, the actual exposure of Plant B personnel to radiation increased in 1950 and attained its maximum in 1951 − 113.3 rem/year per each worker in average. Some of the workers received considerably higher radiation doses (see Table 6). In the early years of Plant B operation the highest radiation doses were registered for the main personnel of the technological shops and the analysts of the Central Laboratory. The average values of the summary doses of external gamma-radiation during a period of 1948–1958 were as follows, in rem [45]:

Radiation supervision service	147.6
CMI&T service	185.1
Power service	127.8
Mechanics service	163.9
Sampling workers of the radiochemical lab	144.3
Main personnel of technological shops	170.8–267.7
Logistics officers, including cleaners responsible for desorption operations in production premises	69.7

The first radiation cases were registered as early as 1949. In March 1950, the PGU Board decreed as follows:

- data about the individual exposure of workers to radiation must be inspected every week and appropriate measures for reducing radioactive contamination of workplaces must be taken immediately;
- the staff and structure of the radiation supervision services at Facilities A[11] and B are to be revised;
- violations of the rules, unrelated to emergency situations, are to be severely punished.

At that time people were punished for the incomplete production of plutonium at Facility A, for the wrecking of scheduled plutonium extraction from irradiated slugs at Plant B, and for the untimely delivery of plutonium to the next facility – Plant V.

Many cases of high exposure were related to the severe secrecy regime established at the center. Lack of information about the amounts of plutonium contained in the equipment units, a restricted list of persons informed of its presence at the facility, absence of reliable data about the safe admissible limits of the equipment loading and transportation of plutonium solutions sometimes led to tragic consequences.

Examples given [43] illustrate the tragic high-radiation cases of the rank-and-file workers and managers of Plant B, as well as of outstanding scientists, primarily from the RIAN, who supervised the operation of the center at the initial phase. Among the persons mentioned in the review were operator A. Kuzmin and engineer A.G. Vedushkin. Some of the plant workers (nearly 36–43 % of the total number in 1950–1951) received radiation doses of 100 to 400 rem despite the rules that were in force in 1948–1952 which allowed a yearly radiation dose of not more than 30 rem [46, 47]. B.A. Nikitin, who was a scientific supervisor of the plant start-up and became a director of the RIAN in 1950, died within two years of a radiation sickness. Professor A.P. Ratner died at age 50.

During the early period of plutonium production 2,089 workers of the center were diagnosed for a professional radiation syndrome, 6 thousand received a total radiation dose above 100 rem, including those who received no less than 25 rem within a year. More than 2,000 individuals had a body burden plutonium exceeding the admissible value (40 nCi). According to published data, 17,245 individuals received yearly radiation doses exceeding an admissible value of 25 rem [46, 56]. From 1948 to 1958 the personnel of the first plants of the Plutonium Center worked under unsafe conditions. The long-term observations allowed radiobiologists-physicians to make a rather reliable judgement about the origin of malignant growths found in the organs of servicing personnel during the first years of radiochemical production. Based on the archival data kept in the local departments of the 3rd Chief Directorate of the Public Health Ministry, the frequency of chronic radiation cases among different groups of the Plant B workers in 1948 to 1958 was found to be as follows:

Average (total) gamma-radiation dose throughout the entire operation period	340±5 rem
Maximum dose per year	150±4 rem
Frequency (percentage) of radiation cases rated to the total number of workers in a group	22.5±0.6%

The frequency of radiation disease is substantially lower among individuals exposed to minor doses. Yet, a large number of workers who did not suffer from a radiation disease were exposed to considerable radiation doses exceeding the now existing limit (5 rem/year) and the limit valid at that time (30 rem/year). This extraordinary fact calls for a special study [48].

Table 7 Oncological mortality of Plant B personnel in % of the total number of workers engaged before 1958 [46].

Radiation dose	Plant B	Plants A and B
Total gamma-radiation dose < 100 rem	4.3±0.4	4.8±0.4
Total gamma-radiation dose > 100 rem	8.1±0.6	8.4±0.5
Maximum yearly dose < 25 rem	4.3±0.5	4.0±0.4
Maximum yearly dose > 25 rem	7.7±0.05	7.9±0.5

Table 7 shows that high radiation doses increase oncological mortality. Considering that the annual mortality of adults in the country caused by oncological diseases was 200 cases per 100 thousand, i.e., 6% within 30 years, one can make a conclusion that at a low radiation dose (no more than 100 rem during the period under consideration and ≤25 rem per year) the oncological mortality did not differ from the overall mortality of adults of the country [46]. However, these encouraging conclusions made in [46–48] more than 30 years after the start-up of the plant were unknown in the first years of the Plutonium center operation.

Measures were taken at the center to reduce the overall exposure of personnel initiated by the center and ordered by the PGU and the Public Health Ministry. In 1941–1951 sources of the excess exposure of the plant personnel were analyzed, factors responsible for the overproduction of liquid radwaste dumped into the open drainage system were investigated.

Apart from the personnel engaged in production, residents of the areas adjacent to the center suffered from radiation.

Hazardous Effects of Radwaste Disposal at Natural Water Basins

Although 45 years has passed since the start-up of the center, people still question why the radwaste was disposed to the Techa River, a tributary of the Iset River which flows into the Tobol River. Who was responsible for taking that decision?

The industrial reactor, radiochemical plant, and other facilities of Center No. 817 were located between the Techa and Mishelya rivers. The low-radioactive waste water of the reactors was disposed at the Kyzyl-Tyash Lake (Basin No. 2). Moreover, Plant B radioactive solutions were dumped into the Techa River. Radwastes were disposed above the Kashkarovskiy and Metlinskiy ponds located above the artificial basin No. 10 of the Techa River (Fig. 7) [57, 58]. Thus, radionuclides got into the Techa River that flowed out of the lake not only from the radiochemical plant, but also from the A and AI industrial flow-type uranium-graphite reactors and from three AV-reactors.[12] Yet, it was believed that Plant B alone was responsible for the disposal of nearly 3 million Ci from March 1950 to late 1951 [58]. Although a partial radwaste disposal into the Techa River was specified in the design, the situation was aggravated by emergencies. For instance, because the sorption of plutonium and fission products was not adequately known, no provision was made in the project for the desorption of the equipment and other technological processes resulting in increased amounts of solutions to be disposed. Corrosion leaks in the main equipment units resulted in the dumping of radwaste not only into the storage tanks (complex S) but also into the Techa River. A commission which was instituted by the order of the PGU administration and headed by A.P. Aleksandrov, Corresponding Member of the Academy, Director of the IFP Institute, found that most of the high-level radioactive waste disposed of into the open basins were not specified in the project but were emergency products.

In 1949, when the storage facilities (complex S) were filled up with high-level radwaste, an insoluble problem arose. Much like in the situation which took place at the first production reactor in late 1948, the authorities had to

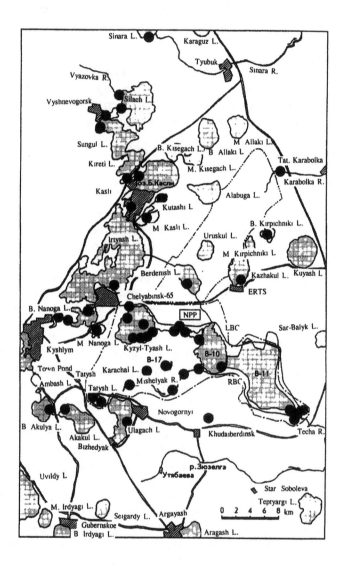

Fig. 5 Location of water basin check points: == – observation zone; –•–•–– sanitary protection zone of the East-Ural fallout area; –•–– sanitary protection zone of the Plutonium Center; ERTS – Experimental Research Test Station; RBC and LBC – right- and left-bank canals; the letter B and a number denotes manmade basins; NPP – South-Uralian nuclear power plant with BN-800 fast neutron reactors.

decide in late 1949 whether to continue the production of plutonium at Plant B or shut it down to stop the disposal of radioactive waste at the Techa River.[13] A decision was taken to continue the plutonium production. Measures aimed at banning the disposal of radioactive water into the Techa River were discussed by a special commission chaired by I.E. Starik, Corresponding Member of the Academy. The commission approved a proposal that was earlier supported by the USSR Public Health Ministry to use the drainless Karachay Lake. The decision of the commission was approved by I.V. Kurchatov, A.P. Aleksandrov, B.G. Muzrukov, and G.V. Mishenkov. Starting from October 28, 1951, most of Plant B radwastes were directed to the Karachay Lake, and merely 100–200 Ci/day were still disposed at the Techa River. Because of the radioactive contamination of the river and the adjacent area 124 thousand residents of the Chelyabinsk and Kurgan districts were exposed to radiation. As many as 28 thousand were exposed to high radiation doses (up to 170 rem). In 935 cases a chronic radiation sickness was diagnosed. Nearly 8 thousand residents were removed from 21 residence areas.

Table 8 presents data on the sickness and oncological mortality rate among exposed and non-exposed residents of the Chelyabinsk region [58]. It was found that the sickness rate and mortality of residents from these districts (including non-exposed individuals) were substantially higher than elsewhere.

For comparison, the number of oncological cases among the people residing in the neighborhood of the Chelyabinsk integrated electro-metallurgical works rated to 100 thousand individuals was as follows [58]:

1985	958
1987	1097
1989	1114

Table 8 Number of malignant growth cases diagnosed in 1955-1982 and oncological mortality (per 100 thousand individuals) among the people exposed to radiation at the Techa River in the vicinity of the Mayak Center [57].

District of Chelya-binsk region	Category	Sick rate	Oncological mortality
Kunashak	Exposed	228.6	144.9
	Nonexposed	158.6	114.1
Krasnoarmeisk	Exposed	319.4	215.4
	Nonexposed	307.9	172.2

In contrast to the operational personnel, the people residing in contaminated areas were subject to an internal exposure. Marrow cells, colon parietes, bone surface, and other tissues were irradiated as a result of the consumption of contaminated water, milk, and other foodstuffs. For instance, researchers from the Institute of Biophysics studied an average marrow exposure for residents from a number of settlements located along the Techa River within a distance of >200 km.[14] Expert opinion is that marrow bone cells are the most sensitive to a radiation damage, and should be considered as a critical organ. For comparison Table 9 presents data on the exposure of various organs of residents from four evacuated settlements and from the Muslyumovo village whose residents were not moved away.

Table 9 Average dose equivalents of various organs and effective doses received by residents of settlements located along the Techa River at different distances from the radiochemical plant.

Settlement	Distance from disposal location, km	Radiation dose equivalent, rem				Effective equivalent dose, rem
		Marrow cells	Bone surface	Colon parietes	Other organs and tissues	
Metlino	7	164	226	140	127	140
Techa-Brod	18	127	148	119	115	119
Asanovo	27	127	190	104	90	100
Nadyrovo	48	95	180	62	44	56
Muslyumovo	78	61	143	29	12	24

More than half (73%) of the highly exposed residents living along the Techa River (28 thousand individuals) received effective dose equivalents of <20 rem, about 12% above 50 rem, and some 8% above 100 rem. Some of the residents might have received 300–400 rem as did some of the plant operators.

In the initial operation period of the radiochemical plant the average effective dose equivalent was 32 rem for the residents of the Chelyabinsk region (Techa areas) and 7 rem for the residents of the Kurgan region.

Later the disposal of liquid radioactive waste to the Techa River was stopped as a result of appropriate measures. However, after the disposal of 76 million m^3 of waste water with a total activity of 2.75 MCi that took place in 1949–1952, the silts of the Techa River still contain a considerable amount of long-lived radionuclides (^{90}Sr and ^{137}Cs). Over the past 30 years the content of these radionuclides has diminished by half as a result of decay. As declared by a number of commissions, the silts of the Techa River (in its upper reaches) should be classified as a solid radioactive waste with all ensuing requirements to their disposal. All of these wastes are concentrated in the protection zone of the center, which includes man-made water basins 10 and 11 (see Fig. 5). Table 10 presents the contents of ^{90}Sr and ^{137}Cs in the water and bottom sediments of the water basins located in the protection zone.

The radiation situation is now monitored by the Experimental Research Test Station and the Environment Protection Laboratory of the Center. The construction of the South-Ural nuclear power plant near basin 10 is in progress.

It is expected that the use of low contaminated water of basins 10 and 11 by this plant will promote a partial rehabilitation of the area [59].

Table 10   Contamination of water basins in the protection zone of the Mayak production association, Ci/l [58].

Water basin no.	Area, km^2	Volume, mln m^3	In water		In silt	
			^{90}Sr	^{137}Cs	^{90}Sr	^{137}Cs
2	19	83	$1.1 \cdot 10^{-8}$	$4.5 \cdot 10^{-9}$	$1.3 \cdot 10^{-6}$	$5 \cdot 10^{-5}$
6	3.6	17.5	$3.7 \cdot 10^{-10}$	$2.0 \cdot 10^{-11}$	$3 \cdot 10^{-7}$	$3.9 \cdot 10^{-5}$
9	0.25	0.4	$1.7 \cdot 10^{-3}$	$1.2 \cdot 10^{-2}$	0.3	1,4
10	16.6	76	$3.5 \cdot 10^{-7}$	$8.6 \cdot 10^{-9}$	$3.5 \cdot 10^{-6}$	$1.5 \cdot 10^{-1}$
11	44	217	$5.1 \cdot 10^{-8}$	$2 \cdot 10^{-11}$	$1.3 \cdot 10^{-6}$	$1.3 \cdot 10^{-7}$
17	0.17	0.33	$7 \cdot 10^{-4}$	$4 \cdot 10^{-6}$	0.12	$3.3 \cdot 10^{-2}$

Related to the context of this paper are the consequences of a radiation accident that took place at Plant B on September 29, 1957, when one of the tanks containing radioactive waste exploded. Of the 20 MCi of radionuclides contained in the tank, 18 MCi were deposited within the limits of the production area, and 2 MCi were dispersed over the area of the Chelyabinsk and Sverdlovsk Districts, known as the East Uralian fallout area of about 1000 km^2. An area of 200 km^2, including the East-Uralian State Reserve, was declared by a governmental decree as a protected zone. An area of 49.3 km^2 (see Fig. 5) was allotted to the Experimental Research Test Station. The inspection conducted after the accident revealed that about 260,000 individuals residing in the contaminated area had received radiation doses exceeding the admissible yearly limit. More than 10,000 residents were moved from the most contaminated sites. More than 5,000 workers who were in the production area were subject to a one-shot exposure as high as 100 rem within a few hours between the moment of explosion and the beginning of evacuation [57]. In the course of remediation of the accident effects, in 1957–1959, nearly 30 thousand workers of the center, civil engineering builders, and military engineering troops received radiation doses above 25 rem.

During the first 10 years of the development of plutonium production at the Mayak Center, the selfless labor of its workers was accompanied by the overirradiation of the servicing personnel and the residents of the towns and villages adjacent to the protection zone. Significant areas and many water basins of the Techa River drainage system need rehabilitation. Some of those people lost their lives in the name of the secure defence of their motherland having involuntarily participated in the armaments race. Though the radiochemical Plant B was shut down long ago, the hazardous effects of its operation will be felt for a long time over the most of the South Ural area.

NOTES

1 A thousandth fraction of a microgram.
2 In a paper published in 1993 Meshcheryakov [51] emphasized a decisive contribution of Khlopin in the development of a program for the production of fission materials and the construction of the first USSR particle accelerator.
3 Later, to reduce the number of waste storage facilities, the Institute of Physical Chemistry (IFKh) recommended the concentration waste radioactivity by evaporating solutions [53]. This technology was not developed and could not be realized in practice.
4 The construction of Plant B started in December 1946.
5 During the first year of Plant B operation the "cooling" time for irradiated uranium did not exceed 45 days.
6 In late 1946 the pipe was planned to be 130 m high (decision taken by NTS Section 4).
7 In the initial period all of the workers participating in the construction and installation operations were provided by the *Glavpromstroy* Department of the USSR NKVD Commissariat.
8 Since 1947 Akimov headed a Corrosion Control Commission under the USSR Academy of Sciences.
9 B.P. Nikolskiy, a physico- and radiochemist, worked at the Leningrad State University and at the RIAN. He was elected a member of the Academy in 1968.
10 I.V. Petryanov-Sokolov is a physicochemist who developed methods for studying aerosols. He was elected a member of the Academy in 1966. Worked at the Karpov Physicochemical Institute since 1929.
11 In 1949 the average dose of the personnel exposure at the industrial reactor was nearly twice that of Plant B; the annual value was as high as 93.6 rem/year.
12 Reactor AV-1 was put into operation on July 15, 1950; AV-2 on April 6, 1951; AI on December 22, 1951; AV-3 on September 15, 1952.
13 According to existing method for calculating a maximum admissible limit of radwastes to be disposed at the river drainage systems, an allowable disposal of fissile products into the Techa River was assessed approximately at 1000 Ci/day.
14 According to a governmental decree issued in 1958, a 200-km area located along the Techa River was declared as a protection zone. The protection zone around the man-made water basins located there measured 134 km^2 in the area.

CHAPTER 6

First Plant for Producing Nuclear Charges (Plant V)

Probably one of the most dramatic pages in the history of nuclear weapons was connected with the development of plutonium metallurgy. High-purity plutonium metal was needed to produce a nuclear charge. Location of a site for this high-security enterprise was selected by the Special Committee early in 1947. This committee (I.V. Kurchatov was among its members) was headed by S.N. Kruglov, Minister of Internal Affairs. The committee agreed upon the construction of the plant on the site of the ex-Navy ammunition depot located near the Tatysh railway junction, not far from the town of Kyshtym [41]. Within two years, in February 1949, Plant V was put into full-scale operation.

The development of plutonium metallurgy and casting technology, as well as research into the characteristics of plutonium and its alloys, are discussed in detail in memoirs by V.S. Yemeliyanov [11, 59]. Kurchatov and Yemeliyanov offered Academician I.I. Chernyaev, Director of the Institute of General and Inorganic Chemistry, who was one of the most prominent experts in handling low-quantity substances (metals of the platinum group), to start research into plutonium characteristics. They promised to provide him with a small amount of plutonium — a bead 0.5 mm in diameter (metallurgists call such a metal bead "regulus") [59]. No one knew at that time how metal plutonium looked, what its melting point was, if it was brittle or plastic, and what physical and chemical properties it had. Infinitesimal amounts of this metal were produced in a cyclotron. Later, when the F-1 reactor was put into operation, studies of plutonium properties started at NII-9 — a basic PGU technological institute[1]. First milligrams of plutonium were produced by NII-9 in 1947 (Facility No. 5) from uranium slugs irradiated in the F-1 reactor. It was only at that time that metallurgists started to work with "real" plutonium.

**ANDREY ANATOLIEVICH
BOCHVAR,**
Academician from 1946.
**Head of the NII-9 Department,
scientific coordinator of the pluto-
nium production plant of Center
No. 817. Director of NII-9 from
1952 to 1984.**

Production technology for Plant V was
developed under the scientific supervision of
the NII-9 experts.

The first production line of Plant V was
intended to produce plutonium for the first
Soviet atomic bomb. The second line was
being constructed for the production of ^{235}U
components.

Besides the PGU managers and Kurchatov, scientific coordinator of the atomic
project, plutonium components intended for
the first Soviet atomic bomb were ordered by
Yu.B. Khariton, Scientific Director of KB-
11.

In 1947, a special Department V headed
by A.A. Bochvar was set up at the NII-9 to
concentrate efforts on plutonium and uranium metallurgy. Before joining the NII-9
Bochvar worked for the Research Institute of
Non-Ferrous Metals and Gold and the Research Institute of Metallurgy (USSR Academy of Sciences). He developed a theory of
plastic material production, and in 1946 was
elected Academician by the USSR Academy
of Sciences. Later he worked at NII-9 first
as the head of a department and then as
director of the institute. His fields of research were crystallization and casting
properties of metals and alloys, recrystallization and heat resistance of metals,
and physical metallurgy of uranium and plutonium. Department V consisted of
three laboratories:

- Laboratory of Radiochemistry, headed by I.I. Chernyaev, Director of the
  Institute of General and Inorganic Chemistry;
- Laboratory of Metallurgy, which concentrated on metal plutonium
  production, headed by Professor A.N. Volskiy;
- Laboratory of Physical Metallurgy and Metal-Working, headed by
  Professor A.S. Zaimovskiy.

As early as August 1947, based on the first experiments conducted on
Facility No. 5 and other data, the NII-9 Design Bureau and researchers from
Department V developed a technology for a refining plant designed for the
purification of a plutonium concentrate (end product of radiochemical Plant V),
which consisted primarily of plutonium and lanthanum fluorides. The proposed
technology was based on a combination of a few different purification

techniques applied in production of a high-purity plutonium. A draft target plan was submitted to the PGU for approval in September 1947; since that time NII-9 started the development of this project.

Plutonium was produced at Facility No. 5 until mid-1948. Uranium slugs irradiated in the F-1 reactor contained only micrograms of plutonium, which were distributed among different research laboratories to study the physical and chemical properties of plutonium, an element that does not occur in nature. Its high chemical and radiation danger was discovered much later. Larger amounts of plutonium were known in its compounds with other elements, in hydroxides, fluorides, nitrates, chlorides, and oxalates.

Chernyaev's laboratory was engaged in the development of two technologies for Plant V: plutonium/lanthanum fluoride concentrate purification, and the purification of nitric acid solution containing sodium/uranium and sodium/plutonium biacetates.

Problems that were faced by other research laboratories of Department V were primarily related to studies of metallic plutonium properties and chemical interaction with other elements. The veil of top secrecy

ILIYA ILIICH CHERNYAEV, Academician from 1943. Director of the Institute of General and Inorganic Chemistry, USSR Academy of Sciences, from 1941 to 1996. One of scientific coordinators of the plutonium production project at the chemical and metallurgical plant of the Plutonium Center, South Urals (1949).

created extra difficulties. As mentioned in [60, 61], the first communication on the achievements of Soviet researchers in the study of plutonium chemical reactions with other elements was made by S.T. Konobyeevskiy, a NII-9 researcher, one of the most prominent figures in metal physics, Corresponding Member of the USSR Academy of Sciences, at the Academy Special Session held on July 1–5, 1955. The pure plutonium was found to be a metal with a melting point of 640°C and a boiling point of 3,227°C. Its structure and physical properties were very much different from those of many metals. Within a range of ambient to boiling temperature plutonium passed through six allotropic modifications (see Table 11). Very unusual was the fact that metallic plutonium density depended greatly on the temperature: at temperatures of 310–450°C it was about 14.7 g/cm^3, and from ambient temperature to 120°C − 19.82 g/cm^3. With this density variation, it was clear that no homogeneous (fracture-free) plutonium could be produced from its melt.

Due to its α-activity, compact plutonium was subject to self-heating: the kinetic energy of α-particles was transformed to heat inside a metal lump. It was found that plutonium gives off heat − 1.923×10^{-3} W/g; therefore a plutonium lump of 50 g must have a surface temperature 5–10°C above the ambient. Lots of yarns have been told about plutonium's self-heating ability. Yu.B. Khariton and Yu.N. Smirnov recollect that in 1949, while showing nickel-plated plutonium hemispheres to a group of KGB and Special Committee top brass in Chelyabinsk-40, A.P. Aleksandrov explained to them that "it was only plutonium that could be so hot" [62]. Renowned columnist V. Gubarev in one of his articles [63] presented his readers with the "hot fact" that early in 1949 Kurchatov, Zernov and Khariton brought to Stalin " ...a glossy sphere of pure plutonium about 10 centimeters in diameter... Stalin cautiously touched it with his palm:

– Oh, yes, it's warm. Is it always warm?

– Always, Iosif Vissarionovich. ...Stalin gave his consent to start tests. In May of 1949 Kurchatov left for the test site."[2]

Note that until May 1949 that amount of pure plutonium was simply nonexistent in the Soviet Union, because a sphere 10 cm across should contain about 10 kg of high-grade plutonium. Early in 1949 Plant V just started to master its production technology. Moreover, in certain conditions a 8 kg sphere of high-grade plutonium might even collapse as prompt neutrons attained a critical mass, i.e., it could be scattered due to a neutron burst [64].

Table 11

Phase	Crystalline structure	Temperature range, °C	Phase transition temperature, °C	Volume change at phase transition, °C	Density, g/cm^3
α	Monoclinic	<119	122	11.0	19.82
β	Monoclinic, space-centered	119–218	206	3.5	17.70
γ	Rhombic, face-centered	218–310	319	7.0	14.70*
δ	Cubic, face-centered	310–450	451	−0.5	15.92
η	Tetragonal, space-centered	450–472	476	−3.0	16.00
$\varkappa$	Cubic, centered	472–640	640	–	16.52

* 17.14 g/cm^3 at 235°C [61].

It was found that plutonium was prone to severe corrosion and formed aerosols quite easily. Due to its high toxicity plutonium was extremely dangerous. Therefore, it had to be handled using glove boxes or purpose-built sealed chambers. All these facts became known later, and at that time safety measures while handling the plutonium at the NII-9 laboratories were rather poor. The same safety problems arose during initial stages of work at various technological facilities and workshops of the newly-built Plutonium Center (Facilities A, B, V) in the Southern Urals. Luckily, only insignificant amounts of plutonium and fission products totaling a few curies were processed at the NII-9 facilities. On the contrary, Plants B and V handled integrated radioactive inventory of tens of thousands curies in solutions, insoluble precipitates, wastes, and manufactured plutonium components.

The low-temperature crystalline forms of plutonium (α-, β- and γ-phases) possess a complex crystalline structure of low symmetry. These phases have a distinctive anisotropy of a thermal expansion factor. For example, this factor is equal to $+84.3 \times 10^{-6}$ $°C^{-1}$ for γ-phase plutonium along one of the crystal axes, and $-19.7 \times 10^{-6}°C^{-1}$. Therefore, on cooling its melt has the form of a powder instead of an ingot.

ANTON NIKOLAEVICH VOLSKIY,
Academician from 1960.
Head of the metallurgical laboratory at NII-9; Deputy Director from 1960 to 1966. Coordinator of the project for production of plutonium metal and uranium-235.

Six allotropic modifications of the metal (below its melting point), substantial volumetric changes and high reactivity demanded a sophisticated technology of producing metal plutonium components. Researchers, who investigated the properties of metal plutonium and its alloys, noted that handling plutonium in conventional conditions was virtually impossible. Soviet metallurgists had to develop special techniques and procedures to provide for safe metal plutonium handling to produce even small amounts (tens or hundreds of milligrams) of homogeneous plutonium alloys [61].

The workers of Department V, A.N. Volskiy, S. Trekhsvyatskiy, V. Sokolov, Ya. Sterlin, and other metallurgists, were able to produce the first plutonium beads just before Plant V was put into operation. Metallographers from A.S. Zaimovskiy's laboratory and researchers from other NII-9 labs carried out intensive studies of the samples [60, 61]. Because of the low ductility of the α- and β-phases of metal plutonium, volumetric changes taking place on cooling led to an internal stress buildup and microcracking, which lowered the density of the metal. Powdered metal and plutonium chips are highly pyrophoric in the air at ambient temperature [61]. Because of the high refractoriness and high reactivity of metal plutonium, casting molds had to be produced of rare and expensive materials. Most suitable were tantalum, tungsten, calcium oxides or fluorides, magnesium and cerium oxides. While casting − melting and teeming − plutonium oxidation was to be reduced to a minimum, i.e., fine vacuum within processing equipment and melters was to be secured. Another difficulty was to cool plutonium castings without their cracking [61]. Also, serious problems were encountered while shaping plutonium castings under pressure. At low temperatures (α-phase) plutonium is very difficult to shape because of its high brittleness. Only when reaching δ-phase (310–450 °C) does plutonium become plastic and, thus, can be easily shaped by pressing, forging, molding, stretching, etc. On cooling, however, plutonium pieces undergo three dangerous regions of phase transformation when plutonium density changes. Therefore, the problems of warping and cracking remain. Both casting and high-temperature pressure shaping of plutonium pieces require a deep vacuum or inert atmosphere. Also special materials should be selected for pressing auger production.

Soon it became obvious that physical properties of metallic plutonium make it extremely difficult to use pure metal. Therefore, the most important problem then was plutonium alloying and doping, as well as studies of its intermetallic compounds. It was found that the most valuable doping compounds were elements that enabled fixing plastic δ-phase plutonium at room temperature [60, 61].

Nowadays, we possess thorough and detailed data on plutonium alloys, same as for basic technical alloys. At those times the bulk of these data was derived from foreign literature. For example, 15 of the 60 phase diagrams in [61] were borrowed from foreign sources: data on plutonium alloys with cobalt, zinc, gallium, cadmium, and other metals. Doping compounds, along with neutron capture, reduce plutonium density. This may lead to a substantial increase of the critical mass and, hence, of plutonium mass in the final products. For example, for a ^{239}Pu sphere of 19.25 g/cm^3 surrounded by a beryllium layer of 5.22 cm, the critical mass would be 5.43 kg [64]. At the same time, for a ^{239}Pu sphere of 15.8 g/cm^3 surrounded by a beryllium layer of 5 cm, the critical mass would be 7.48 kg. For a sphere of 19.25-g/cm^3 plutonium surrounded by beryllium neutron reflectors of different thickness, critical mass can vary more substantially:

Beryllium neutron reflector					
Thickness, cm	5.22	8.17	13.00	21.00	32.00
Plutonium critical mass, kg	5.43	4.66	3.93	3.22	2.47

This had to be taken into consideration not only while selecting a design of a nuclear charge, but also at all stages of metal plutonium processing, handling, and storage.

At the radiochemical Plant B, where irradiated naturally enriched uranium was processed[3], the final products (free of uranium and fission products) had a plutonium concentration high enough to start spontaneous nuclear reaction. The technology used at Plant V possessed an intrinsic hazard of a spontaneous reaction at practically all processing stages. Therefore, the nuclear safety problem for the plant was closely connected, first of all, with plutonium concentration monitoring in every piece of processing equipment, vessels or canisters, pipelines or a smelter. Also, it had to be taken into account that process vessels might contain plutonium precipitates, solutions, etc. Critical masses of fission materials vary at different processing stages. Because of the low critical masses, the smelting process at Plant V was organized using small-size equipment and small amounts of plutonium. Similar problems were encountered at Plant V while handling solutions and precipitates that accumulated in the processing equipment and also radioactive wastes: at that time no experimental data on a minimum critical mass of plutonium were available (only calculated data were at hand). It was only in 1951 that they determined that under certain conditions the critical mass of plutonium in 20–40 g/l solutions was slightly more than 500 grams. Inaccurate laboratory analyses, instrumental errors, or just carelessness of personnel were the factors that might result in exceeding the safety exposure limit of 100–150 g set at that time. Therefore, the total concentration of plutonium in each process vessel and pipeline had to be monitored, taking into account the plutonium precipitation. At that time it was an enormously difficult task: safety levels were often exceeded, especially when handling plutonium solutions, and, therefore, spontaneous chain reactions occurred rather often.

The minimum experimental critical mass in the spherical vessel with a neutron reflector, which contained 38.4 g/l of plutonium nitrate was 510 grams of pure plutonium. These amounts of plutonium were often accumulated at Plant B, and especially at Plant V, where plutonium solution was often held in spherically-shaped vessels while stirring, mixing, etc. The problem might have been aggravated by precipitation, because equipment flushing was not performed on a regular basis.

Additional Requirements to Plant V Products

One of the main theoretical aspects of an atomic bomb design was the duration of a nuclear reaction that occurs during the explosion. Explosion capacity depends on the number of neutrons produced during a chain reaction. At the same time, the reaction takes some time to develop. The whole process takes millionth, or even hundred-millionth fractions of a second. The designers were confronted with new, unexplored areas of natural science. As Academician Yu.A. Trutnev put it [34]: "We have to deal with physical phenomena which cannot be reproduced in the lab: temperatures of dozens or hundreds of million degrees, pressures of millions of atmospheres, density of hundreds of thousand grams per cubic centimeter, and times of one hundred millionth of a second".

When pieces of fissionable material are moved slowly to meet to attain a critical mass nuclear reaction, nuclear explosion does not occur. Instead, a neutron flash takes place, a kind of a powerful burst or a pop with a dispersion of nuclear material. Therefore, an appropriate design of a nuclear charge must be chosen to prevent nuclear explosive from dispersion and, hence, the inevitable breakup of a nuclear reaction. Also conditions must be created to ensure the maximum quantity of fissionable material will react. This is when a nuclear explosion will occur. In 1964 a book [67] by General Leslie Groves, Head of the Manhattan Project in the United States, was published in the USSR. It contained guidelines for technical specifications on parts of a nuclear charge to be used in atomic bombs[4]. (By the way, the leaders of Laboratory No. 2 of the KB-11 design bureau learned these facts back in 1945 from the intelligence sources). The simplest atomic bomb design was based on the use of a so-called "barrel technique" to create a critical mass of fissile material: a subcritical mass of fissile material is directed as a projectile toward another subcritical mass, a target; this leads to the instant formation of a supercritical mass which explodes. This was the explosion principle of the Hiroshima bomb. The first Soviet atomic bomb had the same design. Plutonium hemispheres covered with a corrosion-resistant alloy were manufactured at the Chelyabinsk-40 plant [62]. Documents produced by soviet intelligence and studied by Khariton and Kurchatov contained information on the shape of a plutonium charge, and also described other parts of the US bomb. They described in detail the design of a polonium/beryllium neutron source (initiator) that was inserted into the plutonium charge. The half-life of ^{210}Po is about 140 days, i.e., it is about 4 thousand times more effective than a radium/beryllium neutron source: $T_{1/2}$ for radium is about 1600 years. ^{210}Po was produced in reactor A by irradiation of specially-shaped bismuth blocks:

$$^{209}\text{Bi} + \text{n} \rightarrow\ ^{210}\text{Bi} \xrightarrow{\beta^-}\ ^{210}\text{Po} \xrightarrow{\alpha}.$$

This small-size initiator of a very complicated design contained about 50 Ci of ^{210}Po spread on the beryllium surface.

The generalized report of the NKGB (sent to the Special Committee October 18, 1945, and signed by V.N. Merkulov) stated: "The active material of the atomic bomb is delta-phase plutonium with density of 15.8 g/cm^3. The active element incorporates two hollow-shaped hemispheres, which, like the external spherical initiator, were molded in the atmosphere of nickel carbonyl. The outside diameter of the sphere is 80–90 mm. The weight of the active material (including the initiator) is 7.3–10 kg. ...One of the hemispheres has an orifice

25 mm in diameter used for the introduction of the initiator into the center of the active material, where it is fixed in a special holder... This orifice is plugged by a piece of plutonium." It was not an easy task for Kurchatov, Khariton and the team of physicists, joined by top PGU experts and Plant V research and development coordinator A.A. Bochvar, to accept a technology for plutonium parts manufacturing and their sizes. It was much more difficult to manufacture these parts as prescribed. Nevertheless, this task was done by August, 1949.

As stated in [8, 68], "From the manufacturing plant at Chelyabinsk-40 the plutonium sphere was first brought to Arzamas-16 and then to the Semipalatinsk test site."

Plutonium hemispheres for the first Soviet bomb were covered with a thin nickel film. A special workshop for the task of nickel-plating plutonium parts was established at Plant V. Its first head was A.V. Dubinina. Nickel-plating technology had been developed by researchers of NII-9 and Institute of Physical Problems, USSR Academy of Sciences.

Other types of nuclear charge design were developed in the United States. In his book, General Groves mentions another type of atomic bomb design based on implosion principle: gases produced by the explosion of a conventional high explosive were directed toward fissile material which was compressed until a critical mass was attained. This was the principle used in the construction of a "Fat Fellow", a bomb exploded over Nagasaki. Implosion bomb theory was developed in the United States by S.N. Nieddermeer. He started his research in 1943. The results were used as a basis for the development of an implosion-type plutonium bomb (Fig. 6). This method had great advantages over the barrel technique: the physical properties of plutonium made it difficult to use the metal in a barrel-type bomb [67].

As is seen in Fig. 6, the number, shape, and size of plutonium parts can be modified easily, even with the parts shaped as hollow spheres. This was taken into consideration while selecting equipment and technology for metal plutonium working. In general, an atomic bomb is a highly sophisticated device demanding a high synchronism for the moving plutonium parts. It is obvious that the size of a spherical charge of fissionable material is much smaller than the outer size of the bomb.

The USA was not the only country that developed an implosion technique. An original implosive charge was developed in the Soviet Union by researchers from the KB-11 design bureau: L.V. Altschuller, Ye.I. Zababakhin, Ya.B. Zeldovich, and others. Their charge was successfully tested in 1951. According to Khariton and Smirnov [62], this second nuclear test in the USSR was much more efficient than the first one. The mockups of those two devices − one built using the American technology and the other based on the Soviet technology tested in 1951 − are now on display at the Museum of Nuclear Weapons in Arzamas-16. They are exhibited side by side and show a devastating contrast: the bomb assembled using the Soviet technology is two times lighter than the

American bomb, and at the same time is two times more powerful. Moreover, the Soviet bomb diameter was greatly reduced thanks to an original engineering solution for the implosion process suggested by V.M. Nekrutkin [63].

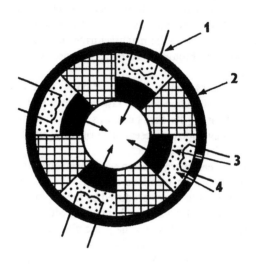

Fig. 6 Schematic of an explosion-type atomic bomb known as a "fat fellow": 1 – electric fuse, 2 – envelope, 3 – nuclear explosive material (^{235}U or ^{239}Pu), 4 – chemical explosive.

The numerous modifications of the nuclear weapons design involved serious changes in the production line of Plant V. Since 1958 schematic diagrams of nuclear and thermonuclear bombs were published in the Soviet literature (Fig. 7), showing more details than the sketch shown in Fig. 6. Every part made of fissile material had a weight (even with a neutron reflector) lower than the critical mass. Total weight of plutonium hemispheres used in the first Soviet plutonium bomb was lower than the critical mass. To increase the fissile material utilization ratio, the charge components were surrounded with neutron reflectors. The joining of the charge parts and the increasing of ^{239}Pu or ^{235}U density was accomplished by the external explosion of a conventional high explosive, initiated by a specially designed fuse [69].

The energy released by a thermonuclear explosion is 80–90% due to a fusion reaction (lithium was used along with deuterium and tritium). A schematic diagram in Fig. 7c shows a three-stage thermonuclear charge, in which explosion energy is emitted sequentially during the fission of plutonium (Stage

1), the fusion of deuterium, tritium, and lithium (Stage 2), and the fission of ^{238}U (Stage 3). Stage 3 starts when thermonuclear neutrons with the energy of about 14 MeV, which is 7 times higher than the fission threshold for ^{238}U, initiate its intensive fission. In this bomb, the bulk of energy is emitted by the process of ^{238}U fission. This principle was substantiated by V.I. Ritus, Yu.A. Romanov et al.[70]. V. Zhuchikhin [68] described in detail the design of the first Soviet atomic bomb including the location of plutonium: an aluminum sphere with a plutonium charge in the middle was placed inside a spherically shaped piece of conventional high explosive.

Fig. 7 Schematics of an atomic (a), a thermonuclear (b), and a nuclear fission-fusion-fission bomb (c) [69]. (a): 1 – neutron source, 2 – conventional explosive, 3 – neutron reflector, 4 – fissionable material, 5 – detonators, 6 – explosion starter; (b): 1 – atomic charge detonator (atomic bomb shown in (a)), 2 – deuterium and tritium mixture; (c): 1 – lithium deuteride, 2 – atomic charge detonator as in (b), 3 – ^{238}U.

Apart from the specific shape, weight, and size of the parts, of great importance were the isotopic composition and purity of plutonium. Appropriate checking techniques were developed by several research centers and approved or not by the Analytical Council of the PGU, which was established on October 26, 1946, and headed by A.P. Vinogradov with P.N. Palei acting as a Secretary. In its Decision of July 14, 1947, the Analytical Council specified a list of permissible impurities in plutonium metal produced by Plant V and approved basic analytical techniques for their control.

In the course of improving the plutonium production technology this list was updated and amended. It is known that impurities are unwanted because they capture neutrons, increase the critical mass, and affect the efficiency of a plutonium charge. Usually in a reactor, a mixture of plutonium isotopes is produced, basically ^{239}Pu, ^{240}Pu, and ^{241}Pu. Depending on its residence time in the reactor, changes occur in plutonium burnup and isotopic composition. Weapons plutonium must contain primarily ^{239}Pu isotope. With a higher nuclear fuel burnup, more ^{240}Pu and ^{241}Pu accumulate in the final product. It is common knowledge that power-grade plutonium (containing 10–20% ^{240}Pu) is not used for nuclear weapons production because the spontaneous fission time of ^{240}Pu is many times smaller than that of ^{239}Pu: too many neutrons are produced and reduce the overall efficiency of the nuclear charge. The half-life during the spontaneous fission of ^{239}Pu is 5.5×10^{15} years, and the neutron yield is 3.2×10^{-2} neutron/(g·s), these values for ^{240}Pu are 1.27×10^{11} years and 1×10^{3} neutron/(g·s), respectively [71]. The content of 1% of ^{240}Pu in the parts made of weapons plutonium creates a neutron background greater than can be expected from the whole charge made of pure ^{239}Pu. Especially unwanted in plutonium parts are light nuclei which react with α-particles emitted by plutonium isotopes, thus increasing the overall neutron background. Quantitative data on neutron yields for light elements such as lithium, beryllium, boron, fluorine, sodium, nitrogen, oxygen, aluminum, silicon, chlorine, etc., can be found in [72].

All of these elements are present in the plutonium production, thus reducing plutonium quality. Concentrations of such elements as boron, beryllium, lithium and fluorine were kept within a range of 10^{-2}–10^{-5}%. The α half-life of ^{240}Pu is 6.6 thousand years (^{239}Pu − 24.4 thousand years); thus it increases the light-metal neutron background by a factor of 4. Therefore, the laboratory monitoring of all process flows during the weapons plutonium production at Plant V was extremely important.

Plant V Designing and Mastering of the Technology

The Plant V project was developed at the NII-9 and IONKh (Research Institute of General and Inorganic Chemistry) and later at the newly formed GSPI-12 Design Institute. Final products of the radiochemical Plant B were used as feed for Plant V. The composition of Plant B end products varied greatly with the improvements of Plant B technology.

A team of NII-9 experts started to work at the Plant V project as employees of the NII-9 Design Bureau which since 1946 incorporated the Leningrad branch of NII-9, previously engaged in designing uranium mines.

According to a governmental decree of February 8, 1948, within the frame of the PGU system, a new design institute GSPI-12 (headed by F.Z. Shiryaev) was set up on the basis of the NII-9 Design Bureau. At first, for secrecy reasons, it was known as a Moscow Design Bureau. Its main task was to design a plant for producing metal plutonium, plutonium components, and also components made of enriched metal ^{235}U. After the initial organizing period, this bureau was headed by A.V. Florov (Deputy Minister of Non-Ferrous Metallurgy) with F.Z. Shiryaev as chief engineer.

As mentioned above, the first line of Plant V was accommodated in the former depot facilities near Tatysh railway junction[5]. That was a separate industrial site, adjacent to the territories of the two basic production facilities of the Plutonium Center located in the area between the Techa and Mishelyak rivers (see Fig. 5). Various documents related to the very first years of Plant V can be found in the PO *Mayak* museum; some of them are cited in [73], a review published on the occasion of the 45th anniversary of the facility. A special decree of March 3, 1948, ordered old barracks (Blocks 4, 8 and 9) to be restructured into a new experimental facility of the plant.

In early 1949, after the overhaul, Block 9 housed an experimental-production facility for processing plutonium solutions which came from Plant B. At midnight on February 26, 1949, the first batch of plutonium solution from Plant B arrived at the chemical treatment shop of Block 9. It was accepted by Ya.A. Filiptsev (Head of the Facility) and I.P. Martynov (Head of the Chemical Treatment Shop) in the presence of B.G. Muzrukov, G.V. Mishenkov, and I.I. Chernyaev. This shop was nicknamed "tumbler facility" by I.I. Chernyaev: the arriving plutonium solution was bottled in platinum containers, which were used for further treatment of the solution. The facility's personnel consisted primarily of women. Among the first "forewomen" of the shop were F.A. Zakharova, A.S. Kostyukova, M.Ya. Trubchaninova and Z.A. Bystrova, graduates of the Gorky or the Saratov University. The metallurgical shop was headed by K.N. Chernyshov [66]. Block 9 was converted to a kind of a chemical lab equipped with wooden exhaust-hoods and a portable equipment: just platinum "tumblers" and settling cones, and golden funnels. All processing stages involved manual labor – no special mechanisms or safety devices for radioactive chemicals handling were available [73].

This is the first impression of M.A. Bazhanov[6], a new employee, on the working conditions at Block 9 [73]: "After passing a checkpoint, I found myself standing in front of a standard barrack building, of which I saw a lot in my life. The barrack was named shop No. 9. My workplace was a tiny room, 5 by 9 meters, with a desk, a chair, and a standard fume cupboard filled with containers, cans and glasses. Behind the cupboard stood metal containers filled with "the product" covered with pieces of plywood." From time to time these containers covered with plywood were used as chairs. These were the working conditions for researchers and processing personnel, the people who developed a technology for the plant. The leading USSR scientists, who were engaged in the technology development, lived in a standard barrack cottage located about 150 meters from the Block 9. Among them were A.A. Bochvar, I.I. Chernyaev, A.S. Zaimovskiy, A.N. Volskiy, A.D. Gelman, V.D. Nikolskiy, V.G. Kuznetsov, L.I. Rusinov and many others.

Plutonium production shop (Block 9).

Construction of new production facilities according to the blueprints produced by the Moscow Design Bureau and the design bureau of NII-9 started in 1948. Facilities were built in two phases. The first phase included shops for the production of metal plutonium from solutions coming from Plant B and casting plutonium articles. The second phase was ready after highly-enriched metal ^{235}U was produced. Part of basic equipment for the rebuilt facilities was delivered to Plant V from NII-9. N.P Fineshin, a veteran of the industry since the setup of the Moscow Design Bureau, recollects that much of the equipment was manufactured at machine-building plants across the country and urgently moved to Plant V site. Some pieces of equipment were manufactured at the repair shop of the Plutonium Center (headed by P.I. Filimonov), in the shops of building contractors, and the shops of the plant itself.

Later, first custom-built workshops with totally different working conditions for the personnel were constructed in accordance with general project designed by GSPI-12.

The plant structure was settled in early 1949. The first director was Z.P. Lysenko (he died soon after the plant was started).[7] P.I. Deryagin (who was previously Chief Engineer of Plant V) became the head of phase 2.

F.M. Brekhovskikh (Chief Engineer of Plant No. 12 till mid-1948, and Deputy Director of Metallurgy at the Plutonium Center) became Chief Engineer. The startup activities of Plant V were supervised by the top managers of the center, B.G. Muzrukov, E.P. Slavskiy, and deputy chief engineer G.V. Mishenkov. The management of the facility was accomplished by the scientists: I.V. Kurchatov, A.A. Bochvar, I.I. Chernyaev, A.N. Volskiy, A.S. Zaimovskiy, L.I. Rusinov, A.P. Aleksandrov, V.I. Kutaitsev, and A.G. Samoilov. Analytical process control was maintained by the staff of the GEOKhI (Research Institute on Geochemistry and Analytical Chemistry).

Special teams of researchers from leading R&D centers across the country came to startup Plant V. These teams consisted of experts from IONKh, NII-9, IFKh, and other research centers in adjacent fields, and were headed by A.N. Volskiy, A.S. Zaimovskiy, A.D. Gelman, and others. They were coordinated by Academicians A.A. Bochvar and I.I. Chernyaev. One of these teams, headed by A.N. Volskiy (Corresponding Member, USSR Academy of Sciences), was engaged in metallurgical research and included a group of young engineers: Dr. V.S. Sokolov, F.G. Reshetnikov[8], I.V. Budaev and Ya.M. Sterlin. A.G. Samoilov[9], a distinguished specialist in plastic working and shaping of metals, was temporarily assigned to the plant and worked under Professor A.S. Zaimovskiy. His group incorporated design engineers M.S. Poido and F.I. Myskov, casting engineer I.D. Nikitin, and young engineers of Plant V B.N. Loskutov and G.M. Nagornyi. This group was to finalize the startup preparation. The pressing equipment developed for the plant was not able to provide smooth and uniform heating of the molded metal (at that moment − a dummy). This worried A.D. Bochvar and A.S. Zaimovskiy. Even more worried were the top management people responsible to the Special Committee: B.L. Vannikov, A.P. Zavenyagin, I.V. Kurchatov, B.G. Muzrukov and E.P. Slavskiy − the August of 1949 was coming. In the shortest time possible the equipment was redesigned and high-quality parts were produced using aluminum dummies. At that time the group of metallurgists headed by A.N. Volskiy produced first metal plutonium in the form of small smelted cylinders. Their mass measured 10% of the mass of two thin-wall plutonium hemispheres that were to be machine-worked after hot molding. This is how A.G. Samoilov described these crucial operations with the amounts of plutonium exceeding a minimum critical mass [38]: "I was ordered to start molding. There were few people in the shop, the physicists placed their instruments around the press and left. Only those responsible for the job remained: A.A. Bochvar, A.S. Zaimovskiy, A.G. Samoilov, M.S. Poido, I.D. Nikitin, and F.I. Myskov. I took the hydraulic lever. The people around were depressed, each thinking agonizingly: will I stay alive, or shall we all disintegrated to atoms? Everybody was thinking: were the physicists' calculations correct? Were they able to consider all factors affecting the increase of critical mass? Can a nuclear explosion be triggered by hot molding? Dead silence fell. Slowly the hob started to go down into the machine, the arm of the manometer started to go up and came to a still at the desired pressure. Molding was finished and heating was turned off. Everybody started to shuffle joyfully, bustle about, and shout. The bosses came in. There were some problems while extracting the part from the mold. At that moment Efim Slavskiy used his huge muscles and extracted the piece from the mold. The part looked shiny and intact. It was then thoroughly and meticulously runstitched using a special attachment to the standard spindle machine. The runstitching process was very scrupulous, consumed a lot of labor, attention, and sconce from the operator. The help of one of our team mates, Mikhail Poido, was indispensable... At the time almost everybody was close to a nervous breakdown – everything seemed unnatural and far away from reality."

The leading shops, laboratories and auxiliary services of Plant V were staffed by PGU and Center personnel, as well as by people from NII-9 and other research organizations. Chiefs of the shops were appointed by a Special Order issued on June 20, 1949: for Shop No. 1 – Ya.A. Filiptsev, for Shop No. 4 – V.S. Zuev, for Shop No. 10 – D.I. Barinov, for Shop No. 10a – N.I. Groshev, and for Shop No. 12 – V.I. Malyshev. I.N. Rozhdestvenskiy was appointed the head of the laboratory.

While handling the first batches of solutions plutonium got smeared, sorbed by precipitates filters, accumulated in flush water, etc. A standard process technology was yet to be developed.

A number of laboratories to control technical specifications for final products were established in Shop No. 11 that dealt with plutonium casting. The laboratories of metal physics, physical metallurgy, x-ray inspection, neutron/isotope physics and a few others were headed by L.I. Rusinov, M.D. Derebizov, V.D. Brodich, V.A. Korobkov, F.P. Butra, V.V. Kalashnikov and G.T. Zalesskiy. In April 1, 1959, the latter became Director of Plant V. From September 23, 1949, to 1959 the Plant was headed by L.A. Alekseev. Each processing stage was supervised by experts from NII-9, IONKh and other research institutes under the control of top people from PGU and the Center. Not only A.A. Bochvar, a scientific supervisor of the Plant, but also top scientific coordinator of the whole Soviet atomic project I.V. Kurchatov and chief designer of the first Soviet atomic bomb Yu.B. Khariton, as well as head of a laboratory at IONKh V.G. Kuznetsov, personally checked the fulfillment of the requirements imposed on plutonium castings by KB-11, NII-19, and IONKh.

In August, 1949, the Center produced first plutonium hemispheres, and the first nuclear charge was successfully tested at the Semipalatinsk test site on August 29, 1949 [54, 74].

It is impossible to mention all people who were engaged in plutonium metal production startup and development: I.G. Evsikov, A.S. Nikiforov, G.T. Zalesskiy, V.V. Myasnikov, S.I. Biryukov, G.M. Nagornyi, K.T. Vasilenko, I.S. Golovnin, I.V. Mikheev, G.A. Aleshina, L.P. Sokhina, Z.A. Issaeva,

F.P. Kondrashova, S.M. Frolov, B.N. Serikov, E.D. Vandysheva, Z.A. Byst-rova, I.I. Oshchepkov, D.A. Olonichev, D.I. Vinogradov, M.I. Gribanova and many others who worked with the Center, NII-9, GSPI-12 and other research institutes, as well as those involved in engineering and construction work at Plant V.

Today, new generation often asks questions regarding the moral philosophy of these scientists who were engaged in the research and design of the most deadly weapon in human history. Did they feel any remorse while creating these lethiforous weapons? Journalist A. Artizov asked this question in his last interviews of E.P. Slavskiy, who was among the first initiators of Soviet atomic industry and the minister of atomic industry from 1957 to 1986. The answer was: "I never noticed any remorse feelings among the people. If there had been such, I am sure, we would have not been able to overcome those heaps of difficulties and develop the nuclear weapon potential of this country in such a short time. Researchers were working like mad, fully overwhelmed by their tasks. They needed no blandishing or daunting; they understood that this country was in need of a nuclear shield" [8]. The same feelings were common for most of the engineers, technicians, white-collar workers, and other employees who worked at the top-priority goal of the country, the prime industrial program of that time.

Interior Exposure Effects

The personnel working at Plant B and, especially at Plant V, were subject to a combined radiation exposure: external beta- and gamma-radiation as well as internal radiation of incorporated plutonium, much more than the people working at the production nuclear reactor. Air in the working areas, especially at the very beginning of the plant activity, contained huge amounts of plutonium aerosols, which were inhaled by the personnel.

During the first years of handling plutonium, no one knew a thing about internal irradiation and, hence, no safety measures were taken. It became clear much later, that the presence of plutonium in a human body greatly increases a hazard of lung cancer. Studies that were carried out for 20 years on 2,346 employees (1,832 men and 514 women), who were in close contact with plutonium, helped establish limits for prolonged external plutonium irradiation, and also doses for internal lung irradiation. Professor V.F. Khudyakov *et al.* [75] who worked at the Chelyabinsk-40 Branch of the State Research Institute of Biophysics developed methods for indirect plutonium dose metering. The plant personnel's cumulative integrated dose was found to be 5.03×10^5 rem. In 20 years 45 cases of lung cancer were registered. Spontaneous (expected) lung cancer rate for the urban population of the country (calculated taking into account patients' age and sex) was found to be 32.6. Data published in [75] show that for the groups of the personnel with cumulative irradiation of less than 400 rem the cancer rate was even less than the value calculated using average data for urban population. At the same time the authors note that for cumulative irradiation higher than 400 rem, the actual cancer rate was 2.6 times higher than expected. Within the limits of the models used, there was no obvious relationship between the number of cancer cases and irradiation dose; lung cancer rate depended greatly on the dynamics of dose accumulation.

Also, experimental data were obtained on plutonium accumulation for the residents of Chelyabinsk-40 located not far from the sanitary zone of the plant and adjacent regions [76]. It was found that for the 45 years of the plant operation, the plutonium accumulation rate for the people who lived in the town since 1950, was about 30 times as great as the average global plutonium concentration.[10]

Table 12 compares plutonium concentrations in the human body for the Chelyabinsk-40 residents and people from other regions. These data were based on the postmortem examinations of 60 adult residents of Chelyabinsk-40 and 128 adult inhabitants from seventeen districts of the Gomel Region, who died in 1990–1991. The results revealed that determination of plutonium concentration in a human body was extremely complicated, even for the people who lived near the nuclear facilities.

Table 12 Plutonium concentration in a human body [76].

Region	Plutonium concentration, Bq	No. of samples
Chelyabinsk-40[*]	3.75±1.45	35
Gomel region	0.27±0.15	126
Ufa (1981)	0.11±0.07	5
Western Europe (West Germany, Great Britain, 1981–1982)[**]	0.07±0.12	30

[*] For residents since 1950.
[**] Calculated from Western data.

Taking into account data of the International Committee on Radiology, the authors of [76] analyzed risks of death due to cancer for the habitants of Chelyabinsk-40. Assuming that plutonium accumulation proceeds at a constant rate, the authors showed that the body of a permanent resident who was born in the town and lived there for 70 years would accumulate about 6.5 Bq of plutonium, i.e. effective equivalent lifetime dose would not exceed 0.33 rem. Therefore, 16.5 cases of malignant tumor per 100,000 habitants could be expected. Taking into account the fact that in other regions plutonium accumulation in the human body is much lower, the authors of [77] concluded that correlation between absorbed plutonium concentration and malignant tumor occurrence could not be established. Therefore, chronic radiation sickness and an elevated death rate could be found only in professionals with a long-time radiation exposure histories. This can be traced in the Plutonium Center personnel. Among 2,900 employees (men), who started to work at the Center in 1948–1953, the constantly monitored (total of 99,600 man-years) death rate was 39.9%; among 2,249 men who were hired in 1954–1958 (total of 68.9 man-years monitoring) the death rate was 21.8%.

Out of 11 groups of malignant tumor localization in employees subject to internal and external irradiation, only two groups showed an elevated death rate compared with the country's average for men. These two groups are: "trachea–bronchi–lungs" and "lymphatic system–blood-producing tissues"; these relations were established only for the employees whose cumulative external γ-irradiation dose exceeded 100 cGr.

The above irradiation doses received by the personnel Plants B and V, as well as of the first production nuclear reactor, especially in 1948–1951, illustrate an irreparable damage to human health, a real human tragedy. Radiation monitoring was poor, and no data on internal irradiation were available at the time. Moreover, during operations with open α-emitters (such as radium, radon, plutonium, and polonium) the effect of internal radiation could not be monitored, and even the term "radiation sickness" was nonexistent. It was years later that maximum concentration limits for these elements were specified. The following allowable concentrations of radionuclides in ambient air were introduced to protect such a critical human organ as lungs: $^{239}Pu - 9 \times 10^{-16}$, $^{226}Ra - 2.59 \times 10^{-14}$, $^{210}Po - 9.3 \times 10^{-14}$ Ci/l. These elements were handled not only at the RIAN, LIPAN and other research facilities, but also at a number of the PGU facilities.

At that time the researchers − physicists and chemists − had only one purpose in hand: to put on stream the first nuclear reactor as soon as possible and obtain long-awaited plutonium, to break the US monopoly on nuclear weapons. Along with the personnel of the Soviet nuclear facilities, there were untimely deaths of a number of top scientists and researchers, fathers and pioneers of the Soviet nuclear industry. Among them were I.V. Kurchatov, V.G. Khlopin, I.E. Starik, B.A. Nikitin, I.I. Chernyaev, A.N. Volskiy, G.V. Mishenkov, A.S. Nikiforov, and many others. Those were the people who, along with their colleagues at research centers and technical personnel of reactor A and Plants B and V, started the plutonium production in the former Soviet Union. It were their efforts, directed by the Special Committee and the PGU, that helped to prevent the Third World War. For over 50 years after the war of 1941–1945 the people of the Soviet Union have been enjoying peace and safety.

NOTES

1 During the early stages of the project NII-9 was the head PGU research institute for uranium production, including development of hydrometallurgical technologies. Later, a separate research organization — NII-10 (or VNIIKhT — Research Institute on Chemical Technologies) — was established to deal with this problem.

2 Yu.B. Khariton denied the fact of visiting Stalin with a plutonium sphere.

3 Later, the plant was converted to process nuclear fuel with a starting enrichment ranging between 2 and 80% ^{235}U.

4 The author [68] referred to a USSR Government Decree signed by Stalin, where the date of testing a plutonium bomb was set for early 1948, and that for a uranium bomb for mid-1948. These dates proved to be unrealistic.

5 These depot facilities were overhauled and renovated in 1947 in accordance with highest standards possible. Special attention (as required by A.A. Bochvar) was put to wall plastering to produce a smooth, "mirror-like" surface to prevent radioactive dust accumulation and provide for easy washing [66].

6 Later, M.A. Bazhanov worked as the head of radiochemical laboratory of Plant B, and further on, headed several radiochemical divisions of Nuclear Reactors Research Institute (NIIAR) in the town of Dimitrovgrad.

7 By a Government Decree (No. 3847-1604) of September 13, 1949, Z.P. Lysenko's family was financially supported: his wife and mother were granted life-long state pensions, and his daughters got educational scholarships.

8 Later, F.G. Reshetnikov, a prominent expert in physical chemistry and metallurgy, became Corresponding Member, USSR Academy of Sciences. He is working at the VNIINM Research Institute since 1946.

9 He worked at the Moscow Center for Hard Alloys until he joined VNIINM Research Institute in 1946, a winner of the Lenin Prize, a four-time winner of the State Prize, an Honored Inventor of Russia, Corresponding Member of the Academy.

10 Global plutonium concentration is the result of cumulative plutonium concentration in the atmosphere produced by nuclear tests in the United States, USSR, Great Britain, France and China.

CHAPTER 7

Concluding Phase of the First Plutonium Bomb Project.
Semipalatinsk Test Site

From the very beginning, Laboratory No. 2 was charged with the production of fissile materials (plutonium and ^{235}U) and the development of an atomic bomb. In April 1946 by a special decree of the Government, the Laboratory's division – Special Design Bureau (KB-11) – was set up to speed up the project. B.L. Vannikov suggested it to be located somewhere not far from one of the existing ammunition production facilities: great amounts of conventional high explosives were needed to continue with the development of an atomic bomb design. Headed by P.N. Goremykin (Deputy Minister for Ammunition Production), a group of A.P. Zernov, Yu.B. Khariton and chief engineer of *Lengorpoekt* I.I. Nikitin (who later was in charge of designing Arzamas-16 facilities) worked for several months trying to find an appropriate location. They selected Sarovsky Monastery named after Venerable Serafim Sarovsky who was canonized by the Russian Orthodox Church in 1903.

KB-11 was headed by the following people: P.M. Zernov – Director from late 1946 to 1950, previously a Deputy Minister for Tanks Production; Yu.B. Khariton – Chief Designer; and K.I. Shchelkin – Deputy Chief Designer.

Appropriate testing facilities were constructed and local production lines – Plants No. 1 and No. 2 – were built. At that time the researchers were studying intelligence reports on atomic bomb design. Research teams for the laboratories were formed. Theoretical Division was headed by Ya.B. Zeldovich.

Top people at KB-11 knew that the US plutonium bomb design was based on the implosion principle: a plutonium charge was squeezed by spherically-focused converging shock waves generated by a conventional explosion [68].

One of the facility veterans, V.I. Zhuchikhin, recollects that "a focusing belt design for a spherical nuclear charge resembled a football: it was an almost regular 32-faceted polyhedron incorporated in a round sphere. The design was developed by V.F. Grechishnikov." The explosion process technology needed to be proved experimentally: a spherically converging shock wave in a high-explosive charge should be provided by special synchronously operating focusing elements initiated by fast fuses. V.I. Zhuchikhin continues [68]: "At first, the sizes of parts forming the core of the charge — plutonium pieces — were selected intuitively rather than searching for optimal parameters, and were adjusted later."

The top security regime at the facility led to situations when even those responsible for the design of the bomb did not know amounts of plutonium needed to produce the bomb, though, as mentioned by Khariton [29, 62], the overall bomb design and shapes of plutonium parts were a more or less common knowledge among those involved in the project. Although very few people knew the amounts of plutonium involved in the bomb design, the top people of the PGU, chief scientific coordinator of the project I.V. Kurchatov, and the chief designer of the bomb knew these data even before KB-11 was formed. That is why even before the possibility of a nuclear chain reaction was proved at Laboratory No. 2 (F-1 reactor), extensive construction of sophisticated facilities for plutonium production and nuclear charge manufacturing was started at Center No. 817 in the South Urals. However, the Soviet researchers, designers, and technologists responsible for the facilities construction were able to pick up just a few things from incomplete intelligence reports. Therefore, enormous efforts were mounted to build a production reactor, develop plutonium production technology, and manufacture at the Center No. 817 final products, nickel-plated metal plutonium hemispheres.

Likewise, great efforts were spent to fulfill several unscheduled jobs at KB-11. In January, 1948, a Special Commission of I.V. Kurchatov, A.P. Zavenyagin, Yu.B. Khariton, A.S. Aleksandrov, and P.M. Zernov submitted at a Special Committee meeting a draft decree of the USSR Council of Ministers "On KB-11 Activities Schedule," where, among other urgent measures, there was an item on the need for experts from other research facilities to be transferred to KB-11 [68]: "After a few amendments the Decree was signed by Stalin on February 8, 1948. As a result, all problems were overcome by August, 1949." In fact, this short concluding remark implied huge efforts to develop:
- automated nuclear charge fusing system;
- radio sensors (jointly with R&D centers and Design Bureaus of the Ministry of Aircraft Production, Ministry of Communication, and other institutions;
- a design for the ballistic casing of the bomb along with a study of its aerodynamics;

– various unique instruments and units.

In his recollections on the development of the Soviet atomic bomb [44] published in late 1992, Yu.B. Khariton writes: " ...the design of the first Soviet atomic bomb was partially based on the detailed schematic and description of the first tested American atomic bomb provided by Klaus Fuchs, and the Soviet intelligence. These reports were available to our scientists in late 1945. When the Arzamas-16 researchers proved that the information was reliable and trustworthy (it required extensive laboratory studies and calculations), the decision was taken to detonate the first Soviet bomb produced in accordance with the proven available American technology." Later in his book Khariton admitted that in that period of severely bad relations between the USSR and the United States and high Soviet experts' responsibilities regarding the success of the first Soviet nuclear blast, any other decision taken at the moment would have been "unallowable" and "light-minded". Everything was top secret at the time: intelligence information, the decision to build an "American-style" bomb, and the list of research people allowed to use American data. He proceeds [38]: "Now I can understand the feelings of the veterans of the Soviet atomic project, who considered the first charge (or the schematic of the first charge) to be the achievement of the Soviet scientists and engineers." It should be noted that a decision to proceed with American atomic bomb technology held back the development of the Soviet design which later proved to be more effective. It took two more years to finalize our own design: the second Soviet nuclear test took place in 1951.

Work on the American and the original Soviet design called for sophisticated calculations and extremely dangerous experiments that were carried out at KB-11. How and by whom these experiments were carried out was described in detail in the memoirs of the Arzamas-16 facility veterans [66, 78].

To organize experimental work and accommodate KB-11, along with Plant No. 550, additional premises were built, and some premises of the Sarovsky Monastery were used [79]. New laboratory blocks were constructed, monastery buildings were restructured, and a dwelling area was built, which later became the town of Arzamas-16. At present, the old belfry of the Monastery is used as a TV tower, and the old building of the Monastery's Refectory was restructured to become a local theater (Fig. 8).

Taking into account the security regime of KB-11, a restricted area was set up around the town of Sarov and the monastery, its perimeter measuring about 50 kilometers. On July 17, 1947, the town of Sarov was removed from the administrative management of the Mordovian Autonomous Republic and excluded from all regional records. All those who lived and worked in this town were strictly prohibited even to mention the historical name of the place; the prohibition was in force for almost 40 years.

The program of structural materials impact compressibility studies started at the facility under K.I. Shchelkin. For this purpose techniques of pressure

Fig. 8 Monastery buildings in Sarov before KB-11 was created.

measurements (from 0 to 2–5 million atmospheres) beyond the front of detonation and shock waves had to be developed. This task was assigned to the newly-established KB-11 laboratories headed by A.F. Belyaev, M.Ya. Vasiliev, L.V. Altshuller and V.A. Zuckerman. Instruments and laboratory units were developed for measuring velocities of converging detonation and shock waves. During laboratory simulations, 10–20-kilogram charges of explosives were blasted. In 1947 first laboratory studies which incorporated high explosive detonation with a simultaneous blast parameters registration were carried out at Vassiliev's laboratory by a group of researchers headed by A.D. Zakharenkov.[1] By that time the atomic bomb schematic diagram was set up and the bomb sizes roughly estimated; KB-11 was engaged in detailing individual elements and parts of the charge. The core of the bomb was a spherically-shaped charge simultaneously initiated at 32 points of its surface. The upper layer of the spherical charge consisted of focusing elements which transformed 32 diverging detonation waves into one spherically-converging wave [68]. An aluminum sphere with a plutonium charge inside was to be placed inside a spherically-shaped explosive charge. Part of the focusing element explosive consisted of trinitrotoluene/hexogene (1:1) alloy with a detonation speed of 7,560 m/s. The other part was made of barium nitrate and some additives to provide a lower detonation speed of 5,200 m/s. This was done because the operation of the focusing element was based on different detonation velocities of its parts. This design of the focusing element provided an equal time for detonation to pass from the initiation point to any point on its inner surface in spite of different paths. The greater the difference in detonation speeds, the smaller is the element [68].

Custom-made instruments and laboratory equipment were designed at KB-11 and IKhF to measure and monitor these parameters, including the asphericity of the detonation front. The optimization of the explosive components ratio was supported by physical modeling and testing. This required an urgent manufacture of explosive cylinders 30–40 mm in diameter and 100 mm in height. Blasting was organized at Site No. 2, not far from a bunker that was built in 1947 5 kilometers from the Sarov plant. Even before the first metal plutonium articles were produced at Plant V of Center No. 817, in April 1949, KB-11 finalized the design of a focusing layer for the first nuclear charge. More and more new effects were discovered during these experiments. For example, E.I. Zababakhin[2] showed that detonation speed must increase as spherically-converging detonation waves approach the convergence point. This was verified experimentally modifying the charge radius [68].

YULIY BORISOVICH KHARITON,
Academician from 1953.
Chief designer of the KB-11 (VNIIEF) from 1946; till 1993 – Scientific Coordinator of the Russian Federal Nuclear Center, Arzamas-16, where Soviet nuclear and thermonuclear weapons were designed. Honorary Scientific Director of the Center till 1996.

By mid-1948 the activities of the KB-11 divisions and technical services were focused on two major objectives. Captain V.I. Alferov was in charge of blasting technique and triggering system development; also, he headed research on the bomb initiation control system. Major General N.L. Dukhov was in charge of a team that developed a high-explosive charge and an overall bomb design; later he was put in charge of the Design Bureau. By that time the main KB-11 theoreticians arrived, first Ya.B. Zeldovich and E.I. Zababakhin, and later – E.A. Negin and G.M. Gandelman.

The most serious attention was given to measuring the speed of combustion products beyond the detonation wave front. These studies were carried out by a laboratory headed by E.K. Zamoiskiy.[3] The instantaneous symmetrical compression of a plutonium charge inside a conventional explosive charge is the basic condition for the creation of plutonium supercriticality which determines the efficiency of a nuclear explosion. The same principle applies to nuclear

BORIS LVOVICH VANNIKOV,
Deputy Director of the Special
Committee (Deputy to L.P. Ber-
iya). From 1945 – Director of the
First Chief Directorate (PGU) of
the USSR Council of Ministers,
Director of the executive body
charged with the creation of the
atomic industry.

charges produced of other types of fissile materials, i.e., enriched ^{235}U and ^{233}U (produced from ^{232}Th).

The efficiency of a nuclear explosion depends on the speed of supercritical mass formation. The critical mass of a fissile material can be surpassed by combining the subcritical parts of a charge or by increasing the density of the charge material by subjecting it to extremely high pressure. The decrease of a critical mass with increasing plutonium density is a well-known fact.

The critical mass of a fissile material (^{239}Pu, ^{235}U, and ^{233}U) is inversely related to its squared density, and the critical volume, to its cubed density [64]. These relations (especially, a density increase with increasing pressure) were not known at that time. However, theoretical estimation of nuclear charge size and weight required the knowledge of a lot of cross-sections of neutron interaction with various materials. Moreover, it was necessary to develop theoretical methods for determining the critical mass of a plutonium charge, i.e., the critical mass of a prompt neutron reactor which, contrary to the F-1 reactor, was not fitted with a moderator.

In the late 1940s three special groups of researchers were set up to deal with mathematical problems. They were headed by K.A. Semendyaev, A.N. Tikhonov, and N.N. Meyman. Later, these groups were united into a new Moscow research institute headed by Academician M.V. Keldysh [80].

Measurements of nuclear constants needed for critical mass calculations were performed by physical laboratories headed by N.A. Protopopov, G.N. Flerov, D.P. Shirshov and Yu.A. Zysin. In mid-1948 these laboratories were able to move to a newly built block. At that time three new laboratory buildings were constructed at KB-11, two three-storied and one single-storied, which were to accommodate divisions engaged in laboratory studies. The number of employees was growing permanently, as did the number of research projects.

The decision to copy the US design of the bomb enabled the Soviet physicists and designers of the first plutonium charge (N.L. Dukhov, V.F. Grechishnikov, D.A. Fishman, N.A. Terletskiy, and P.A. Essin) to avoid, during the early period, the problems and accidents that occurred in Los Alamos during the assembling of the charge and the determination of critical masses for plutonium hemispheres [64]. The KB-11 physicists G.N. Flerov, D.P. Shirshov, Yu.A. Zysin, Yu.S. Zamyatnin, I.A. Kurilov and their colleagues were not able to experiment with critical masses in early 1949, since there was no plutonium available for such experiments at that time. Neither did they have ^{235}U. Therefore, critical masses were determined from theoretical calculations based on the measured cross-sections of neutron interaction with these materials. At a Symposium on the History of the Soviet Atomic Project (Dubna, May 14–16, 1996) Yu.S. Zamyatin spoke about Flerov's experiments made at KB-11, Center No. 817, to assess the criticality of two plutonium hemispheres (fitted with uranium reflectors), which were later used in the first Soviet nuclear bomb [173].

All R&D work at KB-11 was finished by the time when plutonium destined for the first atomic bomb arrived from Chelyabinsk-40. Thus, the Special Committee, PGU managers, and Kurchatov, Director of the Uranium Project, accomplished all preliminary work for the first test of a Soviet nuclear charge at a test site.

Problems with a Neutron Primer

The design of an atomic charge implied that its fissile materials encompass neutron primer, a neutron source used to initiate a nuclear explosion. A special laboratory headed by A.Ya. Apin (later by V.A. Aleksandrovich) was organized at KB-11 to design and manufacture neutron primers. In 1948 it moved to a separate building. In 1949 the laboratory started to produce first primers. The evidence is a short instruction by Yu.B. Khariton that goes as follows [68]: "c/o V.A. Aleksandrovich. Manufacture three sets of neutron primers in June-July."

What activities preceded the manufacture of the first neutron sources? What materials were needed to develop small but highly-intensive neutron sources used as primers? This information, along with some interesting facts on the development of other parts of the first nuclear charge can be found in the memoirs of KB-11 veterans [38]. P.A. Sudoplatov, veteran of the intelligence, notes [174]: "Besides a detailed report from K. Fuchs ("Charles") on the design of an atomic bomb, some data were obtained from Bruno Pontecorvo ("Mlad"). Of much value were data on the design of focusing explosive lenses and the sizes of uranium and plutonium critical masses placed in the explosive device, ...data about ^{240}Pu concentration, a primer, a timetable and a flow sheet specifying the operations of manufacturing and assembling the bomb and a method for initiating its primer."

The intelligence report on the American bomb design contained not only data on the fissile materials used, but also engineering characteristics of the initiator, a neutron primer, which called for setting up the domestic production of an exotic radioactive material such as ^{210}Po. Compared to a radium/beryllium neutron source, where α-particles (He nuclei), produced by ^{226}Ra, bombard a ^{9}Be-target and produce neutrons by the reaction:

$$^4He_2 + {^9Be_4} \quad {^{12}C_6} + {^1n_0},$$

a polonium/beryllium neutron source is thousand times more efficient. The half-life of ^{210}Po is about 140 days compared to that of ^{226}Ra, which is 1,400 years. Therefore, a polonium/beryllium neutron primer could be much smaller in size, but able to produce an intense neutron flux. We knew at that time that ^{210}Po was a product of the natural decay of radium, so it could be extracted from the radium stock kept in the State Reserves Repository of the Ministry of Finance.

Z.V. Ershova [38] recalls that in 1946 Academicians Kurchatov and Khlopin assigned NII-9 to develop a technique for ^{210}Po production from 15 g of radium-equivalent stored at the State Repository. Radium was packaged in 100–150 milligram parcels. Ampoules were under excessive internal pressure because of the presence of helium and radon produced by natural decay during 20 years of storage. This research was headed by Z.V. Ershova, D.M. Ziv, and V.D. Nikolskiy, who knew how to handle radium from their early years with GIREDMET. The group included 15 researchers. Their task was very important and at the same time extremely complicated and dangerous. Moreover, they were very short of time. "We were hurried up but not able to speed up our work, because every wrong move could ruin everything that was done with great efforts... Nowadays, recalling those dangerous operations, it still remains a mystery for me how everyone of us managed not to ruin his health and avoid severe exposure [38]." The ^{210}Po obtained was used for manufacturing polonium/beryllium neutron sources. Because the amount of polonium was too small, the group worked at other types of neutron sources.

The successful production of polonium from 15 grams of impure radium inspired Ershova's group to establish a permanent ^{210}Po production line. In fact, production of substantial amounts of polonium demanded the development of a new processing technology: a recovery of natural bismuth and a manufacture of metal bismuth blocks to be irradiated in the first production reactor by the reaction:

$$^{209}Bi_{83} + {^1n_0} \rightarrow {^{210}Bi_{83}} \xrightarrow{\beta^-} {^{210}Po_{84}} \xrightarrow{\alpha} {^{206}Pb_{82}}.$$

The next step was to develop at NII-9 an extraction technique to separate polonium from bismuth, and at KB-11 to design a new polonium/beryllium neutron source.

Now we know that polonium is as dangerous as plutonium or radium [81]. Because uranium and radium were handled at RIAN, GIREDMET, NII-9, and at several other facilities of the PGU, and plutonium was mainly handled at NII-9 and Center No. 817, polonium/beryllium sources were designed and produced at KB-11, and ^{210}Po, bismuth, and beryllium production was transferred to other facilities.

A program for experimental polonium production was initiated at Laboratory No. 11 of NII-9 [38]. Soon the program started to expand, and the on-stream production was established in 1948 using irradiated bismuth blocks. Until commercial reactor A was put into operation, neutron irradiation of bismuth blocks was performed in experimental reactor F-1 at Laboratory No. 2. Polonium extraction technique consisted of ^{210}Po adsorption from bismuth nitrate solutions by an electrochemically-produced pure powdered metal (bismuth or copper). Polonium concentration in nitrate solution was about 0.01–0.1 Ci/l; using a multistage adsorption/desorption process, the concentration was increased by a factor of millions. This yielded a high purity and specified quality of the final product.

The specific activity of ^{210}Po is 4.5 Ci/mg, the value approximately equal to 10^{13} decays/(min · mg).[4] Taking into account that the chosen neutron primer design demanded the use of 50 Ci, it is clear that the researchers and technical personnel of NII-9 and KB-11 had been worked in extreme conditions.

In the atomic bomb, after the initial blast of chemical explosive, propagating inward through the aluminum layer which surrounds the reflector and through the layer of fissile material (plutonium or ^{235}U), the initiator produces a neutron flux which bombards the plutonium or ^{235}U charge initiating a nuclear explosion. It is clear, therefore, that despite the fact that initial design and specification were obtained from intelligence sources, all of the bomb materials, units, and structures had to be first developed and designed and then manufactured at domestic facilities.

Later, a three-zone layout of production facilities and testing labs was substantiated at Z.V. Ershova's and I.V. Petryanov's laboratories (Karpov Research Institute of Physical Chemistry) to ensure safety for people dealing with polonium. This layout implied the use of special air-conditioning and aerosol removing equipment to reduce polonium discharge well below the approved safety levels.

Special deactivation methods were developed at NII-9 by K.P. Kornilov's group for the surfaces of the premises and equipment contaminated with polonium. The problem of specially-designed high-level ^{210}Po samples was solved in 1948.

Preparation and Testing

Very few people know that in 1946–1949, when work on fissile materials production and a nuclear bomb design was in progress in the USSR, the Soviet researchers had not only intelligence data concerning the design of an atomic bomb. Additional information was obtained in 1946–1947 from other sources. An official Soviet representative, nuclear physicist M.G. Meshcheryakov, was present in mid-1946 at two US nuclear tests at the Bikini Atoll in the Pacific [51].

How did a Soviet physicist happen to attend these US nuclear tests?

It is common knowledge, that after the atomic bombardment of Hiroshima and Nagasaki the international community demanded a ban on nuclear weapons. In his monograph [82] R. Young, a prominent politician and diplomat examined in detail the origin of this international movement for a nuclear weapons ban and analyzed the attitudes that prevailed in the USA and the USSR to atomic weapons development and enhancement. In their desire to show the world the high American potential and power, the US authorities invited two observers from each of the states, members of the UN Security Council, to attend the Pacific nuclear tests. These tests were planned to study the effects of a nuclear explosion on Navy vessels and its impact on the environment. The first charge was exploded in the air above target ships, and the second – underwater. Though the international community strongly opposed these tests, Americans took it quietly just as a test of a new, more powerful and sophisticated weapon. In the United States these two tests of 1946 calmed down the misgivings of the public to the same extent as the first bombs dropped over Japan had aggravated them [83]. Professor S.M. Aleksandrov accompanied M.G. Meshcheryakov to these tests. Captain A.M. Khokhlov, a correspondent of the Soviet Navy daily *Krassnyi Flot* was among the invited journalists.The Soviet representatives observed preparatory activities, targets placement (ships, aircraft, etc.), and the aftermath of the nuclear blasts. In our personal communication on July 13, 1993, which took place in the presence of A.M. Petrossyants, M.G. Meshcheryakov told me that upon his return he had written a comprehensive report (about 110 pages) and handed it over to the Special Committee and the top people from the PGU. Naturally, his data and the film he was able to shoot during the tests were used in the preparation of the Semipalatinsk test site. Later, M.G. Meshcheryakov participated in the organization and preparations of a plutonium bomb test at KB-11 and at the Semipalatinsk test site [54, 68].

One of the most complicated tasks carried out at the KB-11 was the development of an automated system for nuclear charge firing. This system was to meet the following technical specifications [68]:

- explosion must be activated via a cable line 10 km long;
- two control channels must be available to provide for maximum reliability of the system with intersecting electrical circuits at each of its elements. Each signal line was to be equipped with several safety levels;
- electronic modules were to be installed to provide feedback control (to be sure that the control signals were reaching the actuators);
- interlocking devices should be provided for instant blocking of control commands with the system going back to the neutral mode;

- fool-proof cable connectors should be designed and installed.

After all components of the circuit were designed and tested at Arzamas-16, they were field-tested. At Site No. 3 at KB-11 three tests of blasting chemical explosive dummies were performed using aluminum cores instead of fissile material. P.M. Zernov, Yu.B. Khariton, K.I. Shchelkin, N.L. Dukhov, and V.I. Alferov were present at these tests. Thereafter the preliminary stage for the actual test at the Semipalatinsk test site was completed.

According to the technical specifications, the test site was to be located in unpopulated area, 200 km across, near an airfield and a railway station. The site was found 160 km from Semipalatinsk[5]; the area was restricted by the Shagan River (tributary of the Irtysh River) and the Dagilen and Kapyastan mountains [84]. The total area of the test site was 5,200 km^2; its geographical location was 49.7-60.125°N and 77.7-79.1°E. Extensive construction started at the test site prior to tests: technical facilities and living quarters were built. A special settlement was built 120 kilometers from Semipalatinsk on the Irtysh bank, 60 kilometers northeast from the site. The site itself was located in the center of the area. No populated localities were within the test site. The construction took two years to complete; 15 km from the test site temporary quarters were built to accommodate military builders, a small hotel for people attending the tests, a canteen, a boiler and power station, and other auxiliary facilities.

In early June 1949 a State Commission of B.L. Vannikov, I.V. Kurchatov, M.G. Meshcheryakov, A.S. Aleksandrov, and N.I. Pavlov,[6] arrived at Arzamas-16 to evaluate the KB-11 preparation for the first nuclear test. The Commission approved the activities and appointed Yu.B. Khariton and K.I. Shchelkin in charge of organizing the first test.

A State Commission headed by M.G. Pervukhin started in mid-July, 1949, the acceptance testing of the test site facilities:

- control center equipped with shock-wave shelters, explosion control consoles, measuring instruments, and communications equipment;
- building for neutron primer (source) storage, neutron flux measuring equipment adjustment and testing, and design and operation manuals storage;
- buildings accommodating laboratories and shops for plutonium parts storage and assembling, and a workshop;
- targets to be used for explosion evaluation: segments of highways and railroads, complete with bridges and vehicles (automobiles, trucks, railroad cars, etc.);
- segments of underground railway (subway) tunnels 15–30 m deep located 200–300 meters from the explosion epicenter.

Apart from these, two three-storied buildings were constructed spaced 20 m apart (average width of a city street) 800 m from the epicenter to evaluate the impact of reflected shock wave. Different types of fighting vehicles: aircrafts, balloons, tanks, armored carriers, etc., were placed within different ranges from the explosion epicenter. Also, various types of cameras (for normal and high-speed shooting) and measuring instruments for explosion parameters evaluation (shock wave speed, gas cloud development, light emission, neutron flux and γ-radiation) were installed. Several kinds of animals, including two camels, were brought in to study the biological effects of the explosion.

Pervukhin's State Commission finished the inspection of the test site on August 10, 1949. Cameramen made the last shots of the targets: fighting equipment, buildings and structures, underground shelters, and animals.

Special teams of the test attendants were formed to study the aftermath of the first plutonium bomb explosion. Along with people from KB-11, the test was attended by researchers from LIPAN, Center No. 817, RIAN, GEOKhI, IKhF, Optical Research Institute, and from R&D centers of Defense Ministry, Ministry of Public Health, etc. Ya.P. Dokuchaev, who represented Center No. 817 at the test, recollects [54] people from RIAN: B.A. Nikitin, I.E. Starik, B.S. Dzhele-pov, N.A. Vlasov, G.N. Gorshkov, and P.I. Tolmachev. Other organizations, along with people from Ministry of Defense, PGU and the Special Committee, were represented by A.I. Burnazyan, A.P. Vinogradov, M.G. Meshcheryakov, A.P. Aleksandrov, documentary directors and cameramen. A.P. Chelakov, a movie director, who was among those who produced the first documentary, recollects [85] that the number of movie people was strictly limited: "None of our cameramen was allowed to the Riverbank[7] at the time. Their permits were granted later. Practically all shooting was done using unmanned equipment. The weather was gloomy, and no high feelings were among those present. I personally was requested to cut the film of the first plutonium bomb test, which I did...."

Security regime was maintained by MGB people led by Lieutenant-General P.Ya. Meshik. All preparations were finished by August 28, 1949. The chronology of activities in August 27–28 was described in detail in [68]. These are the activities carried out before the explosion on August 28:

- 10.00 to 16.00: charge preparation for arming, plutonium parts and neutron primers carried to the site;
- 16.00 to 21.00: plutonium core preparation – assembly of plutonium articles;
- 21.00 (of the 28th) to 03.00 (of the 29th): plutonium core installation into the charge, neutron background measurements; final focusing element installed, charge cover secured.

Thereafter all measurement systems and instruments were turned on. Photo and movie cameras, as well as oscilloscopes, were loaded. At 04.00 on August 29 the main control console at the Command Post (NP-1) was sealed, and all cable systems were set clear. At 04.30 they started to hoist the charge to the upper platform of the 30-m high testing tower. After the cradle with the charge had reached the top, it was secured to the platform. Procedure of inserting fuses into the charge started at 05.00. This procedures were supervised by

K.I. Shchelkin, the working teams were headed by G.P. Lominskiy and S.N. Matveyev, and their work was scrutinized by two Deputies of the PGU Chief, A.P. Zavenyagin and A.S. Aleksandrov. The people present at the Command Post NP-1 (7 km from the ground zero) during the explosion were L.P. Beriya, M.G. Pervukhin, I.V. Kurchatov, Yu.B. Khariton, G.N. Flerov, some other people from KB-11, and security people. At NP-2, located 10 km south of the ground zero, only 8 persons were present [54].

The reaction of the people in charge of the project, including L.P. Beriya, is described in various publications of KB-11 employees. After mutual greetings and joyful hugs Beriya asked Kurchatov to name the bomb that had been exploded. Kurchatov told him that the name had been already given by K.I. Shchelkin, RDS-1 (abbreviation of the Russian phrase *Rossiya Delaet Sama*, "Russia did it on her own"). Since that time, all of the subsequent nuclear (and hydrogen) bombs were designated RDS-2, RDS-3, etc. [78].[8]

Ten minutes after the explosion, two specially protected heavy KV tanks stuffed with dose metering equipment moved toward epicenter, where the base of the "atomic mushroom" stem had been formed. In his memoirs, A.I. Burnazyan (Deputy Minister of Public Health), who was inside one of the machines, recollects that his tank[9], roaring at high speed, was at the ground zero point in just a dozen minutes after the charge went off. Before returning to report, the crew of the "scout tank" collected readings from the installed meters and fused soil samples. Upon return, the crew reported directly to the Chairman of the State Commission – I.V. Kurchatov, who (along with the Commission members) was at the moment on the road not far from the ground zero point. A.I. Burnazyan recalls that the Commission Chairman listened to the scouts of the radiation safety services with great attention and gratitude. A radioactive cloud produced by the surface explosion was carried away to unpopulated steppe, so the area visited by the Chairman of the State Commission was not too badly contaminated by fission products [11]. The 30-meter tower where the bomb had been placed had disappeared down to its concrete foundation, and a large crater was formed at its place. The tanks were moving over the solidified molten ground terribly crunching under the tank trucks. Everything around was demolished.

Those who were present at the test site described the explosion and its effects differently.

Here are some of the parameters characterizing the explosion at the detonation moment and immediately after it [69]. Within 10^{-6} s, the range of a fireball consisting of sizzling hot vapors and gases produced by the 20-kiloton TNT explosion was ca. 15 m; the temperature in that zone was as high as 300,000°C. Within the next 0.015 s, the range was as great as 100 m, and the temperature dropped down to 5,000–7,000°C. Within 1 s, the fireball achieved a maximum range of 150 m.[10] Because of a high vacuum the fireball rushed upwards at a high speed, and the surface dust was drawn in after it. The fireball cooled down and turned into a swirling cloud of a typical atomic- mushroom shape, observed by those who were present at the test site on August 29, 1949.

Explosion Efficiency and Effects

The main parameters of the atomic charge are the utilization efficiency of fissile material and the yield. These parameters were previously assessed in press as similar to those of the US bomb, probably because of the US design employed in the first Soviet bomb: yield – 20 kilotons TNT and efficiency – 5%. Some of the latest publications report different yields of the plutonium bomb exploded on August 29, 1949 – from 10 to 20 kilotons TNT. Ya.P. Dokuchaev [54] who conducted a radiochemical analysis of plutonium utilization in the first bomb suggests that the explosion yield was even lower.[11]

Explosion of the first Soviet atomic bomb was a surprise to the Western leaders who expected this to occur much later. So, this explosion put an end to the US atomic monopoly.

Because the USA proceeded with the production of fissile materials and were actively testing more sophisticated nuclear weapons, the USSR was forced to increase the production of plutonium and develop a technology for the production of enriched ^{235}U. To this end, the operation of the first commercial reactor at Center No. 817 was permanently improved, and the construction of other nuclear reactors was in progress. On July 15, 1950, an advanced uranium-graphite reactor, AV-1, was put into operation. In 1951, reactors AV-2 (on April 6) and AI (on December 22) were put into operation under the scientific guidance of I.V. Kurchatov and V.S. Fursov, and reactor OK-180 was started up (on November 17) under the guidance of A.I. Alikhanov and V.V. Vladimir-skiy. On May 24, 1952, reactor AV-3 was put into operation. At the first diffusion plant D-1 of another Ural Center (Director A.I. Churin), situated in the Verkh-Neivinskiy settlement, the production of limited quantities of highly enriched ^{235}U was mastered under the scientific guidance of I.K. Kikoin, I.N. Voznesenskiy, and S.L. Sobolev.[12] In order to expand the production of fissile materials, in March 1949 the Soviet Government took a decision to construct a Siberian Chemical Center No. 816 in Tomsk-7 City. Uranium-graphite reactors destined for plutonium production were constructed in Krasnoyarsk-26.

The expanding production of plutonium and ^{235}U called for a sharp increase of the mining and concentration of uranium ore, the manufacturing of more uranium slugs for nuclear reactors, the production of uranium hexafluoride (UF_6) for the diffusion plants in larger quantities, and a further development of radiochemistry and machine- and instrument-building industries. In late 1949, the PGU was reorganized for the purpose of a more effective development of atomic industry and a better coordination of projects involved. A decree issued on December 27, 1949, set up the Second Chief Directorate (VGU), subject to the USSR Council of Ministers, on the basis of the Mining and Metallurgy Department and some subdivisions belonging to other departments of the PGU. P.Ya. Antropov, Deputy Director of the PGU, was appointed Director of the VGU. N.B. Karpov and B.I. Nifontov were appointed his deputies. Besides the major uranium mining Center No. 6 (located in Tadzhikistan), seven enterprises, including a construction organization and Plant No. 48 (*Molniya*) were placed under the control of the Second Chief Directorate (VGU).

However, not only military objectives were placed before the atomic industry. In mid-1950, a Decree issued on July 29, charged the PGU with the scientific, engineering, and organizational management of projects aimed at the utilization of atomic energy in scientific research and for peaceful applications in the national economy. With enriched uranium available in sufficient quantities, various types of research reactors could be built. It was possible to produce considerable quantities of isotopes in special-purpose reactors for the national economy and military purposes. The availability of enriched uranium made it possible to use atomic energy for heat and electricity production. Thus, the epoch of the nuclear power industry began in the USSR.

NOTES

1 Later, from 1968 to 1987 A.D. Zakharenkov worked for the Minsredmash Ministry as a Deputy Minister.

2 Later worked as a chief scientific coordinator at Chelyabinsk-70 Research Center. He was elected Academician in 1968.

3 E.K. Zamoiskiy worked for Laboratory No. 2 since 1947. He developed an electron-optics technique for ultra-high-speed process (10^{-12}–10^{-14} s) chronography. Academician since 1964.

4 The same specific activity value (4.5 Ci/mg) for ^{235}U corresponds to 2.15×10^4 decays/mg, which indicates that uranium could be safely handled in the open.

5 Later this site was extensively used for nuclear testing [84]. In 1961–1989 about 384 underground nuclear charges were tested there.

6 He was responsible for the transportation of plutonium parts of the charge to the test site. He personally signed the waybills upon receiving these parts at the production facility. Later he worked as a Deputy PGU Chief, a Department Director at Minsredmash and a Director of the All-Union Research Institute for Airborne Control Systems.

7 That was the cinema people's nickname for the testing site on the bank of the Irtysh River.

8 Other sources, e.g. [44], claim that RDS-1 designation was proposed by the Secretary of the Special Committee V.A. Makhnev, and that it meant

Reactivnyi Dvigatel Stalina or "Stalin's jet engine". By the way, in the United States they were nicknamed after Stalin — Joe-1, Joe-2, etc.

9 The other tank used another route across the areas with destroyed armored vehicles, bridges, etc.

10 At the detonation moment the temperature in the nuclear reaction zone makes up tens of millions of degrees Celsius thus increasing the pressure up to billions of atmospheres [69].

11 The burnup of 1 kg of plutonium or uranium liberates energy equivalent to 20 kilotons TNT. Even if the efficiency is as much as 10%, the necessary amount of plutonium in the atomic charge must be ca. 10 kg.

12 Initially, the final product consisted of ^{235}U with a mere 75% enrichment.

CHAPTER 8

Experimental and Full-Scale Enriched ^{235}U Production

Unlike plutonium and ^{239}Np[1], discovered only in 1940, uranium has been known since 1789. This element was discovered by German chemist M.H. Klaproth, first in the form of uranium dioxide (UO_2), which for approximately 50 years was taken for a metal. Metallic uranium was obtained for the first time by French researcher E.-M. Péligot in 1841 by reducing uranium tetrachloride with potassium metal [86]. E.-M. Péligot is rightfully credited as the founder of modern uranium chemistry.

For a long time uranium was believed to consist of one isotope with a mass of 238. It was then thought that uranium contained about 0.72% ^{235}U, 99.28% ^{238}U, and traces of ^{234}U. Very soon the prospects of using ^{235}U for attaining a chain reaction and intranuclear energy made uranium the most extensively studied element. The isotopic composition of naturally occurring uranium was revised, and its main nuclear properties were determined. Because the half-life periods of ^{235}U and ^{238}U differ appreciably, about 2 billion years ago the quantity of ^{235}U in the naturally occurring mixture of isotopes was >3%, rather than 0.72%, as nowadays. It turned out that upon enrichment of ^{235}U, equal to the enrichment of uranium in VVER reactors (about 3%), chain nuclear reactions could take place, and that even natural nuclear reactors on thermal neutrons "worked" in the Earth's crust in some rich uranium ore deposits under favorable conditions of humidity and composition of rocks. Such a reactor worked 1.8 billion years ago in Africa, on the territory of present-day Gabon. French scientists established this fact in 1972, when a lower content of ^{235}U was found in the ore from Gabon compared to other deposits [87].

In connection with ^{235}U, the main kind of nuclear fuel, it is worth mentioning that in September 1972, after thorough geological and mineralogical investigations, scientists from the French Academy of Sciences reported at the

General Conference of the IAEA that this natural reactor had worked for ≥ 100 thousand years and generated about 10 GW $\cdot$ year of energy. The burn-out of uranium and the accumulation of plutonium were assessed. In one of the mines of the Oklo Deposit about 500 tons of uranium were found, which in 1972 contained on an average 0.62% rather than 0.72% ^{235}U. A sample of uranium was found, in which the concentration of ^{235}U was 0.296%. The produced ^{239}Pu, whose half-life is 24.4 thousand years, had decayed during that period of time, and was not detected in the samples.

Without dwelling in greater detail on the Oklo phenomenon, it should be pointed out that on June 23–27, 1975 an international conference was held in Gabon, where the Soviet Union was represented by V.A. Pchelkin, Head of the Analytical Service, Deputy Director of the All-Union Research Institute of Chemical Technology. He reported the results of studying 10 uranium ore samples from the Oklo Deposit, collected in early 1975. Products of uranium fission and decay were found in the samples, the concentrations of these products differing from the background concentrations of naturally occurring long-lived isotopes of thorium, lead, plutonium, etc. For example, the quantity of long-lived isotope ^{244}Pu ($T_{1/2} = 8.18 \times 10^7$ years) is millions of times higher than the background values in conventional uranium deposits; the content of ^{232}Th in the uranium samples was comparable with the concentration of ^{235}U. During a period of approximately 100 thousand years of the "natural reactor" operation, ^{235}U not only fissioned there, but also, following a (n, γ) reaction, was transformed to ^{236}U, which then, by way of consecutive decay reactions, transformed into a stable lead isotope:

$$^{236}U \to ^{232}Th \to \ldots \, ^{208}Pb.$$

The half-life period of ^{232}Th is sufficiently long, 1.41×10^{10} years, and the period of its spontaneous decay is $> 10^{21}$; therefore, it has survived almost completely. The half-life period of ^{236}U is 2.342×10^7 years, and during 2 billion years it was completely transformed to ^{232}Th. This means that uranium deposits, where the content of ^{235}U is $< 72\%$, can be found in other regions, where natural reactors had worked.

Contribution of Soviet and German Scientists to the Solution of the Uranium Problem in the USSR

Research into the separation of uranium isotopes in the gaseous phase was started at Laboratory No. 2 immediately after its establishment in 1943. In 1944 L.A. Artsimovich was appointed a head of studies on the method of electromagnetic separation of uranium isotopes.

By mid-1944, the Laboratory of Electrical Phenomena, headed by I.K. Kikoin in the Ural Branch of the Academy of Sciences, was invited to develop isotope separation methods. F.F. Lange, who worked earlier at the Kharkov Physical Engineering Institute (KhFEI) and then developed a gas centrifuge in Ufa, was commissioned to that Laboratory. However, till the end of the war the research was carried out by a small number of workers. Meanwhile, the first 10 g of UF_6, the most stable gaseous compound of uranium, were obtained at the State Union Research Institute-42 (NII-42) of the People's Commissariat of Chemical Industry (at a laboratory headed by B.V. Alekseev). The attitude to the uranium problem changed radically after the end of the war and upon the establishment of the Special Committee and the PGU. German experts, who happened to stay in areas occupied by the Soviet Army, were called to take part in the research [89]. It is well known that in the prewar years and during the war German and American scientists made the major advances in studying the liberation of nuclear energy. The Germans intended to use a so-called "uranium machine" as an engine for rockets and submarines, as well as in the atomic bomb. Soviet specialists succeeded in obtaining data which made it possible to draw a layout of the organization of investigations carried out on this problem in Germany, to ascertain their scope and final results, and to assess an approximate quantity of uranium and heavy water that Germany possessed before the capitulation. These data were reported to B.L. Vannikov and I.V. Kurchatov who directed the research on the atomic problem in the USSR [89]. Therefore, it is not by chance that already in late 1945 a decree of the Council of People's Commissars, dated December 19, 1945, stipulated that German specialists should be called to work in the USSR. A special division headed by A.P. Zavenyagin (9th Directorate)[2], established in the system of the People's Commissariat for Internal Affairs (NKVD) in 1944, had to secure the work of invited German scientists and specialists. Our allies, who after the defeat of Germany gathered in their occupation zone a group of Heisenberg, Gahn, Hartek, Herlach, Strasman, Bote, Diebner, and other scientists, acted in a similar manner. General L. Groves, placed in charge of the Manhattan Project, wrote in his memoirs [89]: "All of the main materials of the Germans were confiscated, but only a few of important scientists were in our hands. At that stage we naturally were anxious that the information and scientists should not fall into the hands of the Soviets."

Both the Allies and the leaders of the Soviet Union, during the teardown of many industrial and research enterprises and other facilities in Germany, connected primarily with the military industry, in some cases offered prominent German specialists to work for the winners under contract with a clear-cut definition of the rights and mutual liabilities. The Soviet party made such proposals to some of the prominent scientists. The offer was accepted by Professor Baron M. Ardenne, head of the laboratory of electronic and ionic physics in Berlin, G. Hertz, Nobel prize winner, head of the Siemens

Laboratory in Berlin, Professors R. Doppel, M. Volmer[3], H. Pose, P. Tissen, Doctors W. Stuze, N. Riel, and other scientists. All in all, about 200 specialists came to the USSR from Germany: 33 doctors of science, 77 engineers, and about 80 assistants and laboratory assistants. By the end of 1948, there were approximately 300 German specialists and qualified workers in the USSR. Some of those who came were engaged in developing a technology for producing highly enriched uranium.

Research into the separation of gaseous mixtures of isotopes were also carried out in the Soviet Union. Under the guidance of the Commission for Isotopes, USSR Academy of Sciences (headed by Academician V.I. Vernadskiy), an All-Union Conference was held in April 1940, which discussed investigations on obtaining heavy water by the method of electrolysis and separation of uranium isotopes from metal vapors by the mass-spectrometric method and by the method of UF_6 thermodiffusion.

As mentioned above, F.F. Lange, an emigrant from Germany, carried out laboratory experiments at the KhFEI using a horizontal high-speed centrifuge for separating uranium isotopes. N.M. Sinev [28] writes that the setup developed by Lange was first transferred to the Sverdlovsk Laboratory of Electrical Phenomena (headed by I.K. Kikoin) in 1944, and then, in May 1945, this Laboratory and Lange himself were moved to Moscow. Some investigations were carried out before the war at other institutes. At the Technical Council of the Special Committee responsible for the implementation of the Atomic Program, already in September 1945, the following reports were made [28]: on September 6, I.K. Kikoin (Laboratory No. 2) and P.L. Kapitsa (Institute of Physical Problems) spoke on "The State of Research for Producing Enriched Uranium by Gaseous Diffusion Method"; on September 10, L.A. Artsimovich (Laboratory No. 2) and A.F. Ioffe (KhFEI) spoke on "Enrichment of Uranium by Electromagnetic Method".

In December 1945, the lines of research were identified and persons in charge were appointed: the gas-diffusion method was developed under the general guidance of Kikoin; the electromagnetic method, under the guidance of Artsimovich; thermodiffusion methods, under the guidance of Aleksandrov and Kikoin. The government specified the scope and entrusted the research and development management in the field of the diffusion method of uranium enrichment to three scientists headed by Kikoin: he was responsible for the physics of the processes, Professor I.N. Voznesenskiy[4] for design and engineering solutions, and Academician S.L. Sobolev[5], for theoretical investigations.

Investigations of the German scientists were intensified. Under the guidance of M. Ardenne, in the premises of the *Sinop* sanatorium in Sukhumi, Institute A was established for developing the following methods [5, 84, 89]:
- electromagnetic method for separating uranium isotopes (research manager M. Ardenne);
- methods for preparing diffusion partitions (research manager P. Tissen);

- molecular methods for separating uranium isotopes (research manager M. Stehenbeck).

In the premises of the *Agudzery* sanatorium (near Sukhumi) Institute G was established, headed by G. Hertz, Nobel prize winner. This institute had to deal with the following lines of research:
- separation of isotopes by the method of diffusion in an inert gas flow (research manager G. Hertz);
- development of a condensation pump (research manager Mülenpford);
- development of a stability and regulation theory for a diffusion cascade (research manager G. Barwich);
- designing a mass spectrometer for determining the isotopic composition of uranium (research manager W. Schütze);
- development of frameless (ceramic) diffusion partitions for filters (research manager R. Reichmann).

The heads of these institutes M. Ardenne and G. Hertz[6] were permitted to invite to the USSR, by their choice, scientists and qualified specialists known to them. They all were given fixed official salaries and were allowed to correspond with and post parcels to their relatives. Soviet physicists from Tbilisi Sh.S. Burdiashvili, I.V. Gverdtsiteli, I.F. Kvartsakho, and others worked together with the German scientists in *Agudzery* and *Sinop*.[7]

German scientists worked also at other objects connected with the use of uranium. Laboratory V, headed by Professor R. Pose, was established in 1946–1947 in the Kaluga District (Obninskoe railway station). From 1946 to 1953 he was one of the scientific supervisors in the development of a nuclear reactor on weakly enriched uranium.

Besides Institutes A and G, and Laboratories B and V, some groups of German scientists worked at Plant No. 12 (N. Riel and P. Tissen), at NII-9 (M. Volmer and R. Doppel), and at Laboratory of Measuring Instruments, USSR Academy of Sciences (I. Schetelmeister). As described in [84, 89], some German scientists were awarded governmental prizes. For instance, Doctor H. Wirtz was twice awarded the Stalin Prize for developing a technology for producing metallic uranium from UF_6. Doctor Stuze was awarded the Stalin Prize for the development of membranes for diffusion machines. Doctor N. Riel was awarded the Stalin Prize and the title of the Hero of Socialist Labor for his contribution to a technology for producing pure metallic uranium.

In 1953, most of the German specialists were relieved of their duties and soon went to Germany. Although the German scientists worked on the general problems solved in the Soviet Union within the framework of the atomic program, they were not connected with the activities of the enterprises and design bureaus located in Arzamas-16 (Kremlev), Chelyabinsk District (Snezhinsk), or Sverdlovsk-45 (Lesnoi) [89]. The secrecy of investigations in separating uranium isotopes resulted in a one-sided estimation of the contribution made by the German scientists to problem of producing highly enriched ^{235}U.

Reminiscences about these investigations were published abroad already in the 1970s, when Soviet specialists were not allowed to publish their works freely.

Foreign authors [84] point out that Professor P. Tissen designed a prototype of a diffusion setup, which was assembled in 1949 at a plant in Verkh-Neivinskiy. He worked in *Sinop* (Institute A) and got a permission to manufacture a diffusion setup which provided 50–90% enrichment of ^{235}U. Later he was involved in the development of a gas centrifuge. In [84] the contribution of the German scientists is obviously overestimated. The contributions of the Soviet and German scientists to solving the problem of producing highly enriched ^{235}U were described in detail in a book by Professor N.M. Sinev, an active founder of this technology [28], and in a review by Professor N.P. Galkin [90], who was engaged at Plant No. 12 in Elektrostal, headed the Central Plant Laboratory in Glazov, headed the Scientific and Engineering Department at the Ministry of Medium Machine Building Industry, and later headed a section at the All-Union Institute of Chemical Technology.

Center No. 813 for Producing ^{235}U

A decree of the Council of People's Commissars of the USSR concerning the construction of the first diffusion plant in the Soviet Union was issued on December 1, 1945, almost immediately after the establishment of the Special Committee. Did a scientific basis for making this decision exist? By that time it was known that the thermal movement rates of the molecules of different chemical elements might differ substantially. This assertion stemmed from fundamental Graham's law (1932), according to which the mean motion speeds of gas molecules at a given temperature depend on their masses. Proceeding from Graham's law, one can suppose that the rates of diffusion of isotopic molecules through a porous partition vary inversely with their masses. Already in 1896 J. Rayleigh, an English physicist, demonstrated that if a mixture of two gases having different atomic weights was passed through a porous partition, the composition of the mixture before and after the partition would be different. In addition to the gaseous diffusion process, some other methods for separating gas mixtures were known. For instance, in 1940 the Commission of the USSR Academy of Sciences, headed by Academician Vernadskiy, discussed plans of research on the separation of uranium isotopes from metal vapors by the method of UF_6 thermodiffusion and by the mass-spectrometric method.

The gaseous diffusion process was known best of all. As early as 1932 G. Hertz succeeded, for the first time in the world, in separating under laboratory conditions a mixture of light noble gases by passing it through porous membranes. He assembled a small cascade of several series-connected separating members equipped with primitive porous membranes. This basic assembly could be used for separating uranium isotopes as well. The decisive factor was that the Soviet government knew from the intelligence data about investigations on producing highly enriched ^{235}U by the gaseous diffusion method. This uranium, as well as plutonium were used to manufacture atomic bombs detonated in August, 1945. Therefore, without waiting for the completion of the investigations of the Soviet scientists and the German scientists invited for a joint research, the Soviet government took a decision to build the first Plant D1, a future Center No. 813, in the settlement of Verkh-Neivinskiy (Middle Ural).

A small difference in the speeds of the isotopes of gaseous ^{235}U and ^{238}U compounds leads to a situation when lighter isotope penetrates the porous membrane faster, and the mixture that passed the partition is enriched with it. The gaseous compound of uranium must be sufficiently volatile and chemically stable, and the chemical element with which uranium is bonded must have only one isotope. Only UF_6 possesses these properties. The pores of diffusion membranes must be small enough for the molecules to pass freely, without colliding with one another, i.e., the diameter of the capillary must be smaller than the free path of the molecules. Getting into these minute pores, the light and heavy molecules of the working gas almost do not collide with each other, but interact only with the walls of the pores. Because the light molecules, containing ^{235}U, are more mobile at the same temperature, a slightly greater quantity of these molecules pass through the pores of the membranes compared to the quantity of the molecules containing the heavier isotope ^{238}U. As a result, more and more light isotopes will accumulate behind the membrane, and the gas mixture of uranium there is enriched in ^{235}U compared with the gas mixture at the entrance to the membrane (Fig. 9).

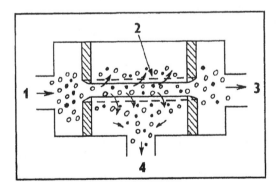

Fig. 9 Diagram of a diffusion process during ^{235}U enrichment: • – ^{235}U, o – ^{238}U. 1 – input of gaseous UF_6; 2 – porous tube (diffusion filter); 3 – uranium depleted in ^{235}U (dump); 4 – uranium enriched in ^{235}U (yield).

ALEKSANDR IVANOVICH CHURIN (1907–1981).
Prominent manager of the atomic industry. From 1946 to 1957 coordinated centers producing highly enriched uranium-235 and plutonium (at Sverdlovsk-44, Chelyabinsk-40, and Tomsk-7). First deputy Minister of the atomic Industry from 1957 to 1970.

The mass numbers of the (heavy and light) UF_6 molecules of the ^{238}U and ^{235}U isotopes are equal to 352 and 349, respectively. Their ratio, equal to 1.0043, defines the ideal (unattainable) coefficient of separation of the mixture of the two uranium isotopes, i.e., the relative concentration of ^{235}U which has diffused through the porous membrane is only 0.0043 times greater than before the membrane. In fact, the passing of UF_6 through one porous partition on one diffusion machine cannot give a separation coefficient higher than 1.002, i.e., the concentration of ^{235}U increases only by 0.2%. In order to obtain enriched ^{235}U of a 90-percent concentration, it is necessary to connect in series tens of thousands of gas-diffusion cells. Requirements to the materials and area of porous membranes and the number of series-connected diffusion machines were known already at that time.

The construction of Center No. 813 (till 1949 known as Plant No. 813 and later as the State Verkh-Neivinskiy Machine Building Plant) started on the basis of Plant No. 261 of the People's Commissariat of the Aviation Industry. In accordance with a governmental decree of December 1, 1945, it was handed over to the PGU by a joint order of the authorities of the PGU and the People's Commissariat of the Aviation Industry, dated December 11, 1945. The first Director of Plant No. 813 was A.I. Churin.[8]

To build the Plant, Building Directorate No. 865 was established by a decree of the Council of Ministers, which started functioning in January, 1946. By June 1946, 6,970 builders worked at the site, and in a year this figure reached 16 thousand. After 1949, up to 25–30 thousand builders of all categories worked at the building site of Center No. 813 [28]. The operations were carried out mostly manually. In the initial period the transport facilities comprised 5 locomotives, 71 automotive vehicles, and 298 horses. First excavators appeared only in 1948. GSPI–11 was appointed as the leading design institute of the Center.

The technology of the first diffusion plant was developed by the personnel of the Laboratory of Measuring Instruments, USSR Academy of Sciences

(LIPAN), and by specialists from Sukhumi. Equipment was designed and manufactured at plants of different industries.

I.K. Kikoin, Corresponding Member of the USSR Academy of Sciences, was appointed a research manager of the diffusion plant; Professor I.N. Voznesenskiy, the former chief designer of the Leningrad Metallurgical Plant, was appointed a deputy research manager. The design calculations for the project were entrusted to Academician S.L. Sobolev, deputy research manager (Deputy Director of the LIPAN). Almost all the work of the scientists and design engineers on the development of the technology for producing highly enriched ²³⁵U was discussed regularly at Section 2 of the Scientific and Technical Council of the PGU, headed by V.A. Malyshev.[9]

In accordance with a governmental decree of December 27, 1945, an Experimental Design Bureau was established at the Leningrad Compressor Plant for developing main separating machines for producing enriched ²³⁵U. E.A. Arkin was appointed a Chief Designer. An Experimental Design Bureau was also established at the Gorky Machine Building Plant early in 1947; it was headed by A.S. Elyan, Hero of Socialist Labor, Director of the Plant: A.M. Savin was appointed a Chief Designer. Both design bureaus worked in accordance with assignments from the LIPAN and its Leningrad Division headed by Corresponding Member of the USSR Academy of Sciences I.N. Voznesenskiy.

ISAAK KONSTANTINOVICH KIKOIN,
Academician from 1953.
Deputy Director of Laboratory No. 2 from 1943 to 1984, Deputy Director of the Kurchatov Institute, Scientific Coordinator of the gas diffusion production of enriched uranium-235 at Sverdlovsk-44.

As pointed out in [28, 89] an Official Report on the development of atomic bomb under the supervision of the US government, translated and published in 1946, proved to be very useful to the Soviet development engineers. In the Report it was stated that out of the four methods of separating uranium isotopes, studied in the USA, preference was given to the gaseous diffusion method.

In early 1946 the Scientific-Technical Council and the heads of the PGU approved the gas-diffusion method for use at the plant being built at the Verkh-Neivinskiy Settlement. This decision did not imply that research into other methods should be abandoned. The work on the electromagnetic method continued under the guidance of L.A. Artsimovich and D.V. Erofeev[10], and on August 21,

1946, the first separation of uranium isotopes on uranium fluoride ions was effected[11]. Progress was made in research into UF_6 separation by diffusion against a stream of steam, which was carried out under the guidance of D.L. Simonenko at the LIPAN (Kikoin's Laboratory). Nevertheless, as N.M. Sinev points out [28]: "H. Smith's Report resolutely and irrevocably supported our attitude in choosing the gaseous diffusion method as a basic one."

What was used to advantage at Plant D-1 from the American publication? H. Smith described, in particular, the main policy in choosing the equipment, the layout of the gas-diffusion plant, and expected difficulties:

- the choice of UF_6 as "working gas" called for the development of vacuum engineering on an unprecedented scale;
- the main difficulties involved with the gas dynamic method were connected with the development of porous membranes and compressor pumps; acres of membranes and thousands of pumps were required;
- a one-stage separation setup (with one compressor) was employed, and therefore several thousand stages were required;
- the best manner of connecting the stages required a multiple reiteration of the cycle: the amount of the material passing through the partitions of the lower stages was many times that of the enriched product separated from the last stage (the stages were connected into cascades);
- at all stages approximately one half of the substance passed through one membrane to the next, higher stage, and the other half returned to the preceding stage;
- the circulation of the total amount of gas at the stage was effected with the help of compressors;
- because the gas flow through each stage varied appreciably with the number of this stage in the cascade, the number and size of compressors also changed from one stage to another;
- the circulation system, consisting of a compressor, a membrane, a pipeline, and valves, was made gas tight for establishing a vacuum.

It was emphasized that neither lubricants nor sealing media or any other material must react with the working gas.

These notes gave an idea of the proper choice of equipment, main requirements to the technology of producing highly enriched ^{235}U, and difficulties that might be encountered by the development engineers, technologists and service personnel in mastering the technology.

The open American publication of the diffusion plant layout in the Official Report, as well as their calculations of the optimal arrangement of the stages [91] suggest that they were sure that other countries were unable to reproduce such a technology of producing highly enriched uranium for the atomic bomb. For instance, the engineering requirements to produce porous membranes for diffusion machines, which follow from the known laws of molecular physics, sounded almost fantastic. In particular, it was pointed out in these requirements

[28] that the diameter of the innumerable quantity of openings must be smaller than 0.1 of the mean free path of the molecules (before collision with other molecules), whose order of magnitude was 0.1 μm. Consequently, the material of the membrane must have millions of openings with a diameter ≤ 0.01 μm and contain almost no openings with a diameter exceeding this value. The openings must not increase in size or become clogged as a result of direct corrosion or dust forming in the case of corrosion somewhere in the system. The membrane must be able to withstand gas pressure of 1 atm, be easy to manufacture in large quantities, and its quality must be homogeneous.

VYACHESLAV ALEKSANDROVICH MALYSHEV, Minister of Atomic Industry from 1953 to 1955. First manager of the industrial production of enriched uranium-235. Chairman of the State Comission for the test of the first USSR thermonuclear bomb.

In the USA serious studies of the diffusion method started in the middle of 1941, and already by the end of that year it was proved that in principle the light and heavy fractions of UF_6 could be separated with the help of a single-stage diffusion setup with porous membranes. In the Report it was pointed out that the filter (membrane) could be made by etching a thin zinc–silver foil with hydrochloric acid. In the preface to the book of H. Smith, Major General Leslie R. Groves, Head of the Manhattan Project, praised the unparalleled technical achievements of the United states. He wrote in mid-1945: "There are no reasons why the history of the administrative and organization measures for making an atomic bomb and the basic scientific conceptions which were central to different practical conclusions could not be made known to the general public. The book contains scientific data, that would not infringe on the interests of the national security." It could not be fully acknowledged that the electromagnetic, centrifuge, and other methods for producing enriched ^{235}U were not promising and had not been used in the USA in the interests of the national security.

Taking into account the difficulties which had to be overcome in developing the gaseous diffusion and other methods for producing highly enriched ^{235}U, one can understand why already in 1945 German specialists were invited to participate in the research (Institutes A and G, located in sanatoriums in Sukhumi.[12] Some of those specialists had a prewar experience in gas diffusion.

In the beginning of 1946 a closed competition was announced for the design of flat filters. Specifications for the development of different filter versions were provided by the LIPAN. Some of the versions were unrealistic. N.M. Sinev writes that there were six versions. These are three of them:
- producing red-copper flat filters by etching zinc in a thin brass plate;
- producing nickel porous plates by punching small needle-like openings on a specially constructed mechanical appliance;
- making porous fabric by a specific arrangement and cementing together of special fibers.

Fifteen organizations, apart from the LIPAN, took part in the competition. A good version was to make a porous plate from fine-dispersed nickel powder, the workpiece being shaped in a press mold arranged on a vibration table, and then sintered. This version was developed at the Moscow Plant for Hard Alloys, Ministry of Nonferrous Alloys, jointly with the LIPAN.[13] Though the parameters of the filters were not very high, they were adopted for production. A technology for preparing fine-dispersed nickel powders was still under development.

Urgent preparations for the development of a technology for producing enriched ^{235}U can be explained mainly by the fact that in 1946 the United States continued to improve nuclear weapons. Two atomic bombs were detonated in 1946 on Bikini Atoll in the Pacific. Moreover, in June 1946, the US Joint Chiefs of Staff completed the development of the first detailed plan of an atomic war against the USSR under the code name "Pincher" [24]. It was planned to deliver a nuclear attack at 20 towns using 50 bombs. Therefore, the Special Committee gave green light to investigations in the Soviet Union, including those carried out by the German scientists, to ensure the soonest possible production of fissile materials.

In addition to the development of flat filters, work was underway in other lines. In 1948 the German scientists at Sukhumi under the guidance of P. Tissen and R. Reichmann in 1948 obtained the first results on the making of two types of tubular filters: framed and ceramic. The first framed filters were manufactured under the guidance of P. Tissen. They were made by applying fine carbonyl nickel powder to a nickel gauze having 7-10 thousand openings per cm^2, followed by sintering in a furnace. Filters of the second type were developed by R. Reichmann and Soviet engineers V.N. Eremin and N.N. Eremina. Ceramic filters were made by extruding a paste-like mass consisting of nickelous oxide (with binders) through an annular die to produce a thin-walled tube, followed by reducing roasting in a hydrogen furnace [28].

Both types of tubular filters, developed mainly under the guidance of the German scientists, after tests at the LIPAN, were approved for use on second-generation diffusion machines. This decision was taken in a dramatic situation, because the opinions of the designers from two design bureaus were diametrically opposite. This problem was considered in detail at Section 2 (Chairman V.A. Malyshev) and at the Scientific and Technical Council of the PGU. It was decided to transfer the Laboratory of P. Tissen from Sukhumi to Plant No. 12 near Moscow; and the laboratory of V.N. Eremin, to the Moscow Plant for Hard Alloys. Commercial production of tubular and flat filters was started at these two plants in 1949.

Many plants from different branches of the industry participated in the execution of Program No. 1, particularly in creating diffusion machinery. For instance, the production of framed filters required manufacturing of a gauze from 0.05 mm thick nickel filaments. Diamond spinnerets were needed for this purpose. The Kolchugino Plant of the Ministry of Nonferrous Metals rapidly mastered a technology for producing finest nickel wire, from which a fine gauze with up to 10 thousand openings per cm^2 was made. Plant No. 12 and the Moscow Plant for Hard Alloys provided the facilities for diffusion ^{235}U enrichment with their super-brittle, top-secret, and extra-pure products. Vacuum pumps and helium leak detectors were developed at the Central Laboratory of the Moscow Vacuum-Tube Works (Director, S.A. Vekshinskiy) and at other enterprises.

The main manufacturers of diffusion machines were the Leningrad Compressor Plant and the Gorky Machine Building Plant. Production of these machines was organized there during a short period of time. Large shops were built for applying galvanic coatings to the inner surfaces of the machines that contacted UF_6. There were hectares of such surfaces, and many thousands of components. The technological process required not only washing and thorough degreasing, but also copper plating in special baths, followed by uniform nickel plating. After these operations all the nickel-coated surfaces were subjected to thorough manual polishing to prevent corrosion in a fluorine medium. "That was infernal, labor-consuming, dirty, monotonous manual work" [28]. Each part, unit, and machine in assembly were subjected to acceptance not only by the quality inspectors of the manufacturers, but also by inspectors of the PGU Checking-and-Acceptance Commission. The Acceptance Commission was headed by Major-General of the Air Force, Professor V.I. Polikovskiy, Director of the Central Institute of Aircraft Motor Engineering, then by L.L. Simonenko (1949–1955), M.D. Millionshchikov, Corresponding Member of the USSR Academy of Sciences[14] (1956–1973), and later by N.M. Lystsov. All of them were workers of the LIPAN.

At the PGU the main responsibility for the coordination of the developments, manufacturing of the equipment, supply of the necessary materials, and control of the terms of building Plant D-1 was placed on the 8th Directorate headed by A.M. Petrosyants. This is what A.M. Petrosyants tells about that time [92]: "The Research Manager and I prepared a large and detailed draft governmental decree, in which we set forth all our requests and demands, and sent it to the Special Committee. In a couple of days I had a telephone call:
–   "Do you really need all these things?!"
–   "Yes, we do", – I answered.

-   "Then come here."

That was a call from the office of the Chairman of the Special Committee, L.P. Beriya. In a private conversation with the author of these lines A.M. Petrosyants told me that, having heard his answers to some questions, Beriya asked with irritation: "Don't you need some pigeon's milk too?" "Well, – I thought, I am going to get it hot, – and answered:

-   "Yes, we do need all of these things."

After some time a question again:

-   "Haven't you overlooked something?"

-   "No, we have taken everything into consideration."

After an active discussion the document, mobilizing a large number of Soviet plants, institutes, design bureaus, and construction organizations, was signed.

The responsibility of the PGU officials, the scientists and the managers of the plants for ensuring the fulfillment of the requests was extraordinary. At the same time, this is indicative of the thoughtful and responsible attitude of the Head of the Special Committee to the successful fulfillment of Program No. 1.

Production of Uranium Hexafluoride

Uranium hexafluoride (UF_6) proved to be the most suitable product: it was a gas which, diffusing through porous membranes, gave an opportunity to separate heavier uranium molecules from lighter ones. An advantage of naturally occurring ^{19}F was that it had no other isotopes. Fluorine is a toxic element, and its maximum tolerable concentration in the air is 1×10^{-4} mg/l. Its reactivity is very high, so that even traces of moisture or lubricants present in the equipment ignite in its atmosphere. The main source of fluorine is fluor spar – fluorite (CaF_2), which contains 48.7% of fluorine. Small quantities of fluorine are present in natural phosphates. In the atomic industry fluorine is used for producing two uranium compounds: tetrafluoride (UF_4) and UF_6 which is the input product of the diffusion production process. Initially, fluorine and its two above-mentioned compounds were produced at the plants of the People's Commissariat of Chemical Industry. As veterans of science point out [90], UF_4 can be produced by treating hexavalent uranium compounds with cheap regents such as hydrogen fluoride and commercial hydrofluoric acid. The reaction between UF_4 and fluorine is a more complicated process which proceeds in several steps. At first the reaction $2UF_4 + F_2 \rightarrow 2UF_5$ gives uranium pentafluoride, and then the reaction $2UF_5 + F_2 \rightarrow 2UF_6$ gives UF_6. The rate of the first reaction is higher than that of the second, and by the moment, when all of the UF_4 has transformed into UF_5, UF_6 either is not formed at all, or is formed in a very little quantity. At room temperature fluorination goes very slowly. In the range of 250–600°C the reaction rate almost does not depend on the temperature. In the process of fluorination (Fig. 10) intermediate fluorides (U_2F_9 and U_4F_{17}) may form [90], and the rate of the process depends not only on temperature but also on the physicochemical characteristics of the starting product. The starting materials for producing UF_6 were not only compounds of naturally occurring uranium, but also the final product of the radiochemical plant, namely, regenerated uranium, in which, after the separation of plutonium, traces of other admixtures might be present along with uranium compounds. Therefore, almost simultaneously with the decision to build the first diffusion plant, a decision was taken to arrange commercial UF_6 production.

The first grams of UF_6 were obtained at Research Institute-42 of the People's Commissariat of Chemical Industry. Fluorination of regenerated uranium was complicated, because the latter contained traces of γ-active isotope ^{232}U. The distribution of radioactive admixtures was studied later under the guidance of Professor E.M. Tsenter [93]. However, in the starting period of the work the content of ^{232}U in the nuclear reactor had to be negligible because of a short uranium irradiation time, and it did not increase the radioactivity of uranium regenerate at all.

The order of M.G. Pervukhin, People's Commissar of Chemical Industry, dated October 8, 1946, prescribed Plant No. 752 in Kirovo-Chepetsk[15] to arrange UF_6 production. In late 1949 the following shops were built at the Plant: for fluorine (Block No. 2), for hydrogen fluoride (Block No. 1), for hydrofluoric acid (Block No. 43), for UF_6, and for the regeneration of waste, as well as stores for raw materials and finished products. A few auxiliary shops were built: refrigeration and oxygen-producing facilities (Block No. 5), a shop for producing UF_4 (Block No. 53), from which UF_6 was derived, and the like.

Uranium hexafluoride is volatile substance. At < 1 atm pressure, i.e., in a low vacuum, and at temperature $< 56°C$, UF_6 is a gas. At lower temperatures the gas goes over to white crystals having a density of about 5.1 g/cm^3. If a cylinder with UF_6 is heated to $> 65°C$ at > 1.5 atm, it can be transferred to the liquid phase [28, 90].

UF_6 reacts actively with water, yielding a mixed oxide UO_2F_2 and hydrofluoric acid. Reactions of UF_6 with organic substances and most of the metals also proceed vigorously. As a UF_6 molecule gives of part of its fluorine, UF_6 immediately passes into a stable solid substance, UF_4, which precipitates onto different surfaces of the equipment in the form of a fine-dispersed green powder. Formation of hydrofluoric acid from the interaction of UF_6 with water and its interaction with any compounds and lubricants required tremendous efforts to develop techniques and equipment for handling UF_6 in the production of enriched uranium. Details of the development of technology for producing and handling UF_6 were described by N.P. Galkin [90]. This veteran was an active developer of the technology for UF_6 production at the plants that were built later in the system of the Medium Machine Building Industry. As mentioned above, the enterprises engaged in the production of UF_6 and other materials were transferred to the Ministry of Atomic Industry from the Ministry of Chemical Industry. When the Kirovo-Chepetsk Plant was transferred to the Ministry of Medium Machine Building Industry, it had the following shops:

No. 1 for producing hydrogen fluoride and hydrofluoric acid;
No. 2 for producing UF_6;
No. 93 for producing UF_4;
No. 110 for producing calcium chloride;
No. 82 for producing chlorine.

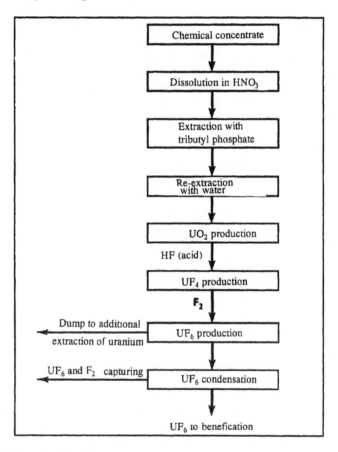

Fig. 10 UF_6 production flowsheet.

Later, a new technology for producing UF_6 by direct uranium oxides fluorination was developed under the guidance of A.S. Leontichuk.

Problems in Starting a Gas-Diffusion Plant at Center No. 813

Plant D-1 was built in the Middle Ural, at the site of an Aviation Industry facility which was transferred to the PGU together with the residential settlement. The main equipment was to be installed in an unfinished block whose area was about 50,000 m^2. The site was conveniently located near a railway station, and there were two deep artificial lakes in the valley of the Neiva River.[16] There was an electric power transmission line nearby. All this was of great importance, because the newly created diffusion production process demanded much power. To produce 1 kg of uranium, 90%-enriched with the ^{235}U isotope, about 600 thousand KW · h of electric power was needed to drive electric motors of the compressors that pumped gas through porous membranes. This process required 175–220 kg of natural uranium [28]. Depending on the degree of ^{235}U extraction, the depleted UF$_6$ flow contained only 0.2–0.3 % ^{235}U and was dumped. These dumps were condensed into the solid phase and delivered in special steel cylinders to storehouses for long-term storage and subsequent utilization.

In accordance with the design assignment issued by the LIPAN and approved by the PGU, in April 1946 GSPI-11 started developing the project. I.Z. Gelfand was appointed the Project Manager (from 1948 the Project Manager was M.M. Vzorov). Leading designers were M.M. Dobvulevich, V.F. Chekalov, I.S. Broido, S.S. Maizel, G.G. Vodopiyanov, and others.

The calculations for a schematic flowsheet for the construction and interaction of 56 cascades composed of thousands of machines of different types and sizes were carried out under the guidance of Academician S.L. Sobolev. The automatic control equipment for all machines and cascades of the plant was developed by the designers of the Experimental Design Bureau of the Leningrad Compressor Plant and manufactured at the Leningrad Compressor Plant both for the diffusion machines of its own design and for the machines delivered by the Gorky Machine Building Plant.

From the very start the diffusion machines were developed in accordance with the design assignment different from the American one-stage design, in which each diffusion cell was served by one compressor.

In early 1946, Kikoin and Voznesenskiy, the Research Managers of the gaseous diffusion method, issued an assignment to the Gorky Machine Building Plant and the Experimental Design Bureau of the Leningrad Compressor Plant for the development of multi-stage machines[17], whose obvious advantage was that dozens of centrifugal compressors, separated by porous membranes, were mounted on one shaft of an electric motor. This machine was a self-contained separating stage of a gas-diffusion cascade. In spite of a large amount of work, in late 1946 the conclusion was made that the idea of a multi-stage machine was erroneous, because the structure proved to be cumbersome, complicated, and difficult to manufacture.[18] In accordance with the decision of the PGU, by the end of 1946 the first lots of 20 one-stage small-capacity machines were manufactured at the two plants. This caused a delay in putting Plant D-1 into operation.

The Experimental Design Bureau of the Gorky Machine Building Plant designed machines coded as OK-7, OK-8, and OK-9 with a mass UF$_6$ flow rate of 30, 90, and 240 g/s, respectively, with a flow pressure in front of porous partitions being 19 mm Hg. The Experimental Design Bureau of the Leningrad Compressor Plant designed machines T-2 and T-15 with similar parameters.

The performance tests of OK-7 and T-15 machines, carried out at the LIPAN demonstrated that the machine designed in Leningrad failed to meet the specifications (tightness, reliability of the bearing supports, etc.). It was decided (Section 2 of the Scientific and Technical Council, Chairman V.A. Malyshev) to furnish Plant D-1 only with GMZ[19] machines manufactured at the Gorky Machine Building Plant. Failures in the development of the multi-stage machine and the T-15 machine, their discussion in the PGU and at Special Committee did not pass without consequences for those who had developed those machines. In June 1947 Professor I.N. Voznesenskiy suddenly died of heart attack; E.A. Arkin was relieved of his duties as the Chief Designer of the Experimental Design Bureau of the Leningrad Compressor Plant, and N.M. Sinev was appointed instead. During the war Sinev had worked as Deputy Chief Designer at the Leningrad Compressor Plant and the Ural Integrated Plant producing heavy tanks. Concurrently, V.M. Malyshev, Head of Section 2, who at that time had been the Minister of Transport Machine Building, "reinforced" the personnel and approved the experimental facilities of the Experimental Design Bureau of the Leningrad Compressor Plant, and transferred it to a round-the-clock operation. E.A. Arkin was appointed Deputy Chief Designer of the Bureau.

Plant D-1 was built up of 6,200 series-connected machines, including: 2,756 OK-7, 2,100 OK-8, and 1,344 OK-9 machine-stages.

The outlying engines were provided with a special ceramic partition which sealed the stator space of the electric motor from the rotor.

The machines were configured into 56 cascades, interconnected by communications for the transfer and takeoff of the enriched gas fraction and the ^{235}U-depleted dump.

All of the equipment was designed to operate continuously; any stoppage might bring about a mixing of UF_6 fractions with different ^{235}U contents. All the cascades were divided into groups, each group consisting of 12 machines. Any groups could be disconnected from the working cascade with the help of vacuum shut-off valves to minimize the mixing of the already separated uranium isotopes during shut-downs and repairs.

The stringent date of launching Plant D-1 necessitated the reconstruction of some large shops at the Leningrad Compressor Plant and Gorky Machine Building Plant and the mass production of diffusion machines even before their commission tests and acceptance for service. In early 1948 railway échelons with OK machines began to arrive to the Verkh-Neivinskiy settlement. Throughout 1948 the State Acceptance Commission carried out accelerated tests of the OK-7, OK-8, and OK-9 machines. This Commission was headed by Professor V.I. Polikovskiy, Director of the Central Institute of Aircraft Motor

Engineering, a prominent expert in aircraft compressors. Though the tests of the machines were incomplete, the Commission issued recommendations for starting their full-scale production. The machines were mounted at Plant D-1 as soon as they arrived there.

Already on May 22, 1948, after the report of the PGU and Research Manager Kikoin, a government resolution was adopted, which authorized submission of the first phase of the Plant for a startup. In compliance with the same resolution, A.I. Churin was appointed the Chief Engineer of Plant D-1, and A.L. Kizima[20] the Director. The former Chief Engineer of Plant D-1 M.P. Rodionov was appointed the Head of Production, and N.M. Sinev, the Chief Designer of the Experimental Design Bureau of the Leningrad Compressor Plant was appointed the Head of the Technical Department of Plant D-1. I.K. Kikoin was appointed Deputy Director for Research. Expert designers engaged at the Leningrad Compressor Plant and Gorky Machine Building Plant were appointed as heads of the leading shops. For instance, G.G. Letemin, an experienced aircraft engineer, became the Head of a Shop for Small Machines (OK-7); N.V. Alavdin (Head of the Test Facilities at the Experimental Design Bureau of the Leningrad Compressor Plant) became the Head of a Shop for Medium Machines (OK-8); and I.I. Afrikantov[21], Head of Pilot Production at the Gorky Machine Building Plant, was appointed as Head of a Shop for Large Machines (OK-9). Naturally, these appointments were not always voluntary.

In 1948, the manning table of the Plant was authorized to be 600 people; 2 years later the personnel consisted of 3,500 people. The engineering and technical personnel of Plant D-1 took an accelerated training course at the LIPAN. The training was conducted I.K. Kikoin and S.L. Sobolev, and also by their colleagues: N.A. Kolokoltsev, Ya.A. Smorodinskiy, A.G. Plotkina, and others.

Back in 1946 a Central Laboratory was established at Plant D-1. Initially it was headed by P.A. Khalileev, and from 1949 to 1953 by I.K. Kikoin. The personnel of the Central Plant Laboratory included experienced specialists transferred from other institutes and plants, and also post-graduates. The Design Bureau of the Plant was headed by V.D. Lourie, the Control Equipment Service and Adjustment Group was headed by M.L. Raikhman, transferred from Laboratory No. 2. B.V. Zhigalovskiy, also transferred from Laboratory No. 2, headed a Group for calculating the startup and transient conditions of newly started cascades. Monitoring of the isotopic composition of uranium was arranged (I.S. Izrailevich and N.A. Shekhovtsev[22]). Special Startup and Adjustment Offices were organized at the Production Board. A.I. Savchuk and P.P. Kharitonov, B.S. Puzhaev and S.A. Kalitin, A.S. Martsiokha and V.P. Maslennikov, V.D. Pushkin and D.M. Levin worked successively at these offices, respectively. Services of the Chief Powerman (A.I. Rybintsev), of the Chief Mechanic (V.N. Osipov), and other subdivisions were set up.

Troubles started at the plant immediately on completion of the startup and adjustment jobs. The diffusion machines (of all three types) developed troubles on reaching the working conditions. Nevertheless, in November 1949 the Plant yielded the first finished products in the form of UF_6 enriched to 75% in ^{235}U. Failures happened with ball bearings and compressor motors, whose rotation speed under working conditions was 6,000 rpm. The bearings broke down after several hundred hours of operation, and some of them rotated normally only a few tens of hours. Replacement of the faulty compressors was started. Work for the replacement and repair of the machines was excruciating, because before an emergency shutdown the machines were filled with chemically aggressive, slightly radioactive UF_6 [28]. However, the radioactivity of UF_6 was not responsible for the failure of bearings in the compressors of the diffusion machines. The tolerances (clearances and interferences) chosen by the designers did not allow for the actual thermal expansion of the components in the bearing pair, that occurring in the case of insufficient heat removal in vacuum. It was decided to replace the bearings in all machines (about 5,500), to correct the tolerances at the machine building plants, and change them in the machines mounted at Plant D-1.

The PGU Commission, headed by the Chief Engineer of the Second Directorate A.A. Zadikyan[23], discovered another cause that interfered with the normal operation of the plant: inadmissibly high corrosion of the equipment, caused by the decomposition of the working gas, UF_6. As a result, high enrichment in ^{235}U was almost unattainable at the final cascades: UF_6 decomposed and turned into a UF_4 powder which precipitated on the walls of the machines.

Leading physical chemists and chemists from the Institutes of the Academy of Sciences, including prominent Soviet scientists such as Academician A.N. Frumkin, Corresponding Member A.P. Vinogradov, Professors I.V. Tananaev, S.V. Karpachev, V.A. Karzhavin, and others, as well as the German physical chemists working in Sukhumi, were invited to work out recommendations for ensuring the operability of Plant D-1. Their visit to the Plant lasted more than a month. Dismantling of many machines revealed that the motors of the compressors had unprotected surfaces (iron stator and rotor laminations), and the high temperature of the motor in the working engine accelerated UF_6 degradation. It was decided to replace the engines in the OK-7 and OK-9 machines, after cleaning the latter from UF_4, by outlying engines, like in the OK-9 machines, to introduce the passivating treatment of the inner surfaces of the equipment, and to lower the temperature of cooling water. On the proposal of P. Tissen and V.A. Karzhavin, the internal surfaces of the machines in all of the cascades and the pipelines were subjected to overall passivation using a heated fluorine-and-air mixture. To minimize the ingress of moist air into the communications, it was decided to build a shop for producing dry air at Plant D-1. The shop with production capacity of 40 thousand m³/h was commissioned a year later. Dry air was supplied through hermetically sealed large-section air ducts provided with bypasses and special shutters. In addition, N.M. Sinev, Head of the Technical Department of the Plant, suggested that the cascades should be supplemented with a terminal OK-6 mini-machine. The proposal was realized without delay. All of these measures soon made it possible to produce highly enriched uranium with a 90% ^{235}U concentration. To replenish the service personnel with specialists, the engineering management of the Plant was reshuffled. Professor M.V. Yakutovich was transferred from the Ural Division of the Academy and appointed Deputy Research Manager of the Plant. M.L. Raikhman, A.S. Martsiokha, N.M. Sagalovich, N.A. Kolokoltsev, M.I. Kalganov, and some other specialists from Laboratory No. 2, who had long been on a business mission at the Plant, were included in its staff. In November 1949 A.L. Kizima was dismissed and A.I. Churin was again appointed the Director of the Plant; M.P. Rodionov[24] was appointed the Chief Engineer.

By 1950, the diffusion technology was mastered in the Soviet Union, and several tens of kilograms of 90% ^{235}U uranium were produced annually [28].

The United States, too, had to overcome difficulties when starting-up their first diffusion plant at Oak Ridge. It is appropriate to recall the words of H. Smith [28] that the group concerned with gaseous diffusion deserved a reward for courage and perseverance, as well as for scientific and technical gifts, more than any other group which participated in the Manhattan Project.

The situation at Plant D-1 was watched very closely by the head officials of PGU and the Special Committee. In the spring of 1949 M.G. Pervukhin stayed at the plant during 3 months without a break. A.M. Petrosyants, a veteran of the Soviet industry, recollects [92] that L.P. Beriya, Chairman of the Special Committee, Member of the Politbyuro, came to the Diffusion Plant in 1949 together with the PGU head officials. Having summoned about twenty leading scientists and plant officials and listened to the reports made by I.K. Kikoin and A.M. Petrosyants, as well as several particular reports about the deadlock with the diffusion process, Beriya interrupted the conference and declared literally the following (no record was taken in shorthand): "The State has given you all you've asked for in excess in spite of great difficulties. Therefore, I give you a three-month term for solving all the problems with starting-up the plant. And I warn you: if you do not succeed, collect pieces of dried bread (an equivalent of "get ready to go to prison"). There will be no mercy. If you do everything as required and succeed we shall reward you." And he left. Mind, he was a man who never used words lightly [92]. The designers and scientists already knew the causes of the failures and did everything they could to solve the problem. Their efforts were crowned with success. Soon L.P. Beriya fulfilled his promise. For starting-up Plant D-1, which ensured the production of the target product with 90% enrichment, a large group of scientists, designers, engineers, technicians, and workers were presented with government awards.

In 1951, for the creation of the diffusion machines and auxiliary equipment and for their full-scale production, Stalin Prizes were awarded to:

- A.S. Elyan, V.D. Maksimenko, A.I. Savin, A.E. Sokolov, K.V. Borodkin, E.N. Chernomyrdik, and A.S. Kalinin from the Gorky Machine Building Plant, and also to V.M. Ryabikov, Deputy Minister of Armament;
- N.M. Sinev, E.A. Arkin, A.I. Zakhariin, V.Ya. Chernyi, Kh.A. Murinson, and N.A. Sorokin from the Leningrad Compressor Plant, and also to V.A. Malyshev, Minister of the Transport Machine Building Industry.

Prizes were given to many members of staff of the LIPAN, of the GSP-11, of Center No. 813, and of other plants and organizations that had taken an active part in starting-up Plant D-1.[25]

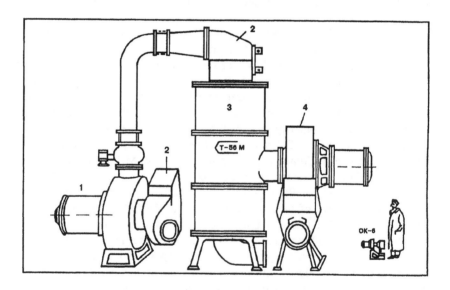

Fig. 11 **The largest and the smallest diffusion machines, T-56M and OK-6, respectively: 1, 4 – full and half flowrate compressors, respectively; 2 – refrigerator; 3 – divider.**

Great advances were made later when building commercial gas-diffusion cascades.[26] An unusually high tightness was attained in the cascade which contained as many as several tens of thousand flange connections. Veterans, Galkin *et al.* [90], illustrated this by the following example: "If one evacuates all of the volumes to a high vacuum, then stops the evacuation and closes the evacuated volumes, then pressure, as is known, will increase gradually because of air leakage. It will take hundreds of years for pressure to reach 0.5 atm". As the technology was gradually mastered, the Plant operation was improved, and the system of regulating the flows of the light and heavy UF_6 fractions in each stage was updated. The problem was complicated by the fact that the flow at the end of each stage was almost one millionth of that passing through the head stage. Therefore, the size of the machines at the end of the cascade was many times smaller than the size of updated machines placed at the beginning of the process (Fig. 11).

One of the parameters characterizing the diffusion machine is the separation work unit (SWU). It is common practice to measure the separating efficiency of a stage in terms of the work done during one year, i.e., SWU/year. The separation work characterizes the consumption of energy for the separation of isotopes, the degree of physical efforts used in the separation of uranium isotopes, and depends on the concentration of the isotopes in the starting material, in the enriched product, and in the dump (tailings). The SWU is directly proportional to the output of the enriched product.

The sizes of gas-diffusion plant are quite impressive. At a comparatively small plant designed to produce daily, say, 1 kg of uranium with a ^{235}U concentration of about 90%, 220 kg of uranium with a ^{235}U content of about 3% are dumped. The total area of filters (porous partitions) at such a plant is as great as many hectares. The total flow of UF_6 is 150 thousand tons per day, this being approximately 100 million times greater than the quantity of the highly enriched ^{235}U thus produced [90]. During the operation of a gas-diffusion plant, a very large quantity of UF_6 must always be present inside the stages and in the auxiliary equipment.

A visitor, who happens to be in the shops of a gas-diffusion plant for the first time, will see machines in its premises, arranged very close to one another and occupying the entire floor areas of the shop (Fig. 12).

After the adjustment of the production process at Plant D-1 and the output of more advanced diffusion machines, Plant D-3 (Director V.D. Novokshenov; Chief Engineer I.S. Parakhnyuk) was put into operation in 1950–1951 at Center No. 813. In 1950 it was decided to build a next plant: Plant D-4 (Chief Engineer A.I. Savchuk; Chief Mechanic I.N. Bortnikov). 2,242 machines developed at the Experimental Design Bureau of the Leningrad Compressor Plant, types T-45, T-46, T-47, and T-49[27], were mounted at Plant D-3. Machines T-45 and T-46, like machines of the OK type, had flat filters; the others had tubular filters. To supply Plant D-3 with electric power, a 75 MW electric power station had to be built.

Plants D-4 and D-5 were equipped with more efficient machines. The characteristics of the diffusion machines that were developed, manufactured, and installed at the plants of the Center for extremely short periods of time are presented in Table 13. The electric power consumed by Plant D-4 which was commissioned in December 1952 (its tail part was ready in the 4th quarter of 1953) was as high as 100 MW. Plant D-5 was designed to accommodate process equipment in blocks, whose total area was 130,000 m^2 or 13 ha.

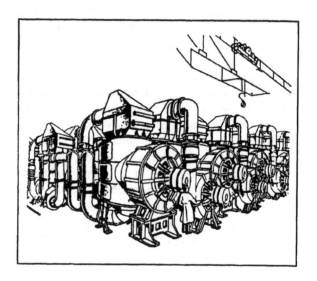

Fig. 12 Cascade connection of T-56 machines at a diffusion plant.

By early 1953, about 15 thousand diffusion machines operated at the plants of Center No. 813. The actually consumed electric power was 250 MW. More than 1 million m^3 of water were daily pumped through the system of the machines of the diffusion plants from the lakes created by the Demidovs on the Neiva River. The round-the-clock operation of the Center with all its production facilities was ensured by 3,500 persons. This is how a new, complicated, high-technology industry was being created in the Soviet Union.

What was the cause of the race in boosting the potentialities of the diffusion plants for producing appreciable quantities of nuclear explosive of the second kind, 90% ^{235}U?

In the book *To Win the Nuclear War: Secret Military Plans of the Pentagon*, published by M. Kaku and D. Axelrod in 1957 in the USA, it was pointed out that immediately after the victory over Japan the US Joint Chiefs of Staff developed detailed plans of an atomic war against the USSR [24]. There were other plans, besides the plan "Pincher". For instance, in accordance with the directive approved by President H. Truman on November 23, 1948, it was planned to drop atomic bombs on 70 towns, including Moscow and Leningrad in 1952 ("Bushwhacker" Plan). These plans were continuously updated. The next plan ("Dropshot", 1949)[28] stipulated dropping 300 atomic bombs on 200 towns in the USSR in 1957. The USA hurried to make atomic bombs for fear that the USSR could catch up with them in the creation of atomic weapons [24].

Table 13 Soviet-made diffusion machines.

Type	Gas flow-rate, kg/s	Gas pressure in front of filters, mm Hg	Power of electric Motors, KW	Mass of stage, tons	Specific power for compressors, KW/(kg · s)	Separation capacity, SWU/year
One-compressor machines with flat filters						
OK-6	0.008	14	0.5	0.2	62.5	0.13
OK-7	0.030	19	1.5	0.7	50.0	0.57
OK-8	0.090	19	4.0	1.2	44.4	1.7
OK-9	0.240	19	10.0	2.5	41.7	3.6
OK-19	0.030	19	1.2	0.66	40	0.6
T-44	0.100	17	3.2	1.1	32	2.0
T-45	0.350	20	14	3.5	40	9.1
Two-compressor machines (boosting-type) with tubular filters						
T-46	0.600	25*	13.5	6	22.5	18
T-47	1.200	52	24.5	6	20.4	45
T-49	2.200	55	44	10	20.0	75
OK-23	4.000	49	60.4	11.5	15.1	140
OK-26	8.000	97	116.3	12	14.5	300
T-51	12.000	90	192	21.3	16	450
T-52	14.000	105	209	20.7	14.9	500
OK-30	14.000	190	198.1	13.5	14.2	500
T-56	25.000	180	369	23	14.8	850

Note. Flat plate-type porous filters.

All of these factors accounted for the intensive building of industrial nuclear reactors and diffusion plants, carried out in the USSR in the postwar period, for producing fissile materials and using them in nuclear weapons. Three more diffusion plants for enriching uranium were built and put into operation in 1949–1964: in Tomsk-7 at the Siberian Chemical Center, in Angarsk – Angarsk Electrolysis Center, and in Krasnoyarsk-45 – Electrochemical Plant [94]. Thus, the industry producing enriched uranium was dispersed for strategic considerations.

Centrifugal and Electromagnetic Isotope Separation Methods

Concurrently with the gaseous diffusion method, other enriched uranium technologies were under development: centrifugal and electromagnetic. After 1945, the German scientists who worked in Sukhumi (Institutes A and G) also participated in this work [28, 84]. Research into the electromagnetic method was carried out under the guidance of M. Ardenne. Professor M. Stehenbeck headed the work on the centrifugal technology.

In early 1951, full-scale production of 90%-enriched ^{235}U was started at the gas-diffusion plant. At that time all other methods were in the state of laboratory research, though the first separation of uranium isotopes had been effected by the electromagnetic method already on October 21, 1946.

German scientists were the first who proposed a centrifugal method. F. Lange, a German emigrant, initially worked at the Kharkov Physical Engineering Institute: he was engaged in developing a horizontal high-speed centrifuge. In 1944 he was transferred to the Laboratory of Electrical Phenomena in Sverdlovsk. Somewhat later, the development of a centrifuge for producing highly enriched uranium was undertaken by Professor Stehenbeck in Sukhumi, wherefrom the project was transferred to the Leningrad Compressor Plant on September 15, 1951. M. Stehenbeck's colleagues were transferred from the Sukhumi Physical Engineering Institute together with him. In Leningrad Stehenbeck worked with his German assistants, H. Zippe and others, and also with several Soviet specialists, including the Head of the Laboratory, engineer A.S. Voznyuk, who had also been transferred from Sukhumi. Later the work on M. Stehenbeck's centrifuge was stopped, and Stehenbeck together with his coworkers was transferred to Kiev to the Academy of Sciences of the Ukrainian SSR. In 1956 they were repatriated to Germany.

The operating principle of the gas centrifuge for separating uranium isotopes is that in the cylindrical rotor of the centrifuge, filled with UF_6, at peripheral velocities >400 m/s in vacuum, heavier molecules concentrate at the rotor wall and descend. The development of the Soviet design of an industrial centrifuge by an enactment of the government in 1951 was entrusted to the Experimental Design Bureau of the Leningrad Compressor Plant. At first the design development was based on Stehenbeck's centrifuge scheme. N.M. Sinev, Chief Designer of the Experimental Design Bureau, notes [28] that in 1953 the Experimental Design Bureau proved that the Sukhumi design of the centrifuge was not suitable for industrial application. Only one unit, a bearing needle unit, on which the basket spins like a top, was used in the Soviet design. This made it possible to solve the most complicated problem of providing high-speed bearings.

The main principle of the design and operation of the Soviet-made centrifuge is displayed on the diagram of Fig. 13. In addition to the bearing needle unit, a specific role in the machine is played by the magnetic suspension, which removes load from the bearing needle, when the basket rotates at high speeds. The rapidly rotating basket can separate light- and heavy-mass gas molecules. An individual gas centrifuge has a relatively small separating capacity. The ideal theoretical separating capacity of a centrifuge is equal to the square of a difference between the molecular weights and to the fourth power of the linear velocity of basket rotation [95]. Because the gas-filling of the centrifuge is low, it needs no special compressors for pumping gas along the communications of the centrifugal cascade. The pressure difference necessary for gas pumping is provided inside the centrifuge due to the gas-dynamic head at the ends of stationary takeoff pipes. Internal circulation of gas in the basket is provided in a similar manner. The main advantage of the centrifuge method over the diffusion method is its small power consumption and substantially higher thermodynamic efficiency. Its power consumption is ten times less than that of the gaseous diffusion technology.

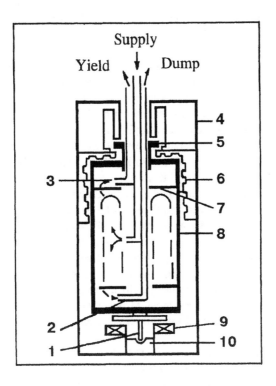

Fig. 13   **Schematic of a gas centrifuge: 1 – needle; 2, 3 – upper and lower withdrawing outlets, respectively; 4 – housing; 5 – magnetic suspension; 6 – molecular pump; 7 – diaphragm; 8 – basket; 9 – stator; 10 – shock absorber.**

The centrifugal technology is particularly effective for separating isotopes of intermediate and heavy elements. To this end, it is necessary to provide gaseous substances in the centrifuge. A specific gaseous chemical compound has to be synthesized for each element to be separated. The uranium centrifuge was created for operating on UF_6. M. Stehenbeck recalled: "My work at centrifuges ended in Leningrad. My several coworkers from Sukhumi and I passed on our experience to a group of physicists, mathematicians and designers already skilled in production matters. Using their own ideas in engineering applications, they soon left behind our results. I do not know, how things went on there after that... " [28]. It is known, however, that H. Zippe, Stehenbeck's colleague, who, while working at the Experimental Design Bureau of the Leningrad Compressor Plant, had access to all data on centrifuges, patented "Zippe's centrifuge" in 1957 in 13 countries: Patent No. 1071597. Actually, this was the centrifuge that had been developed by the specialists of the Experimental Design Bureau of the Leningrad Compressor Plant.[29] Nevertheless, nothing was done in the Soviet Union, when the partial plagiarism became known, because those investigations in the USSR were confidential. In order not to give publicity to the fact that the centrifugal method of producing enriched uranium, more efficient than the gaseous diffusion method, were developed in the Soviet Union, Zippe's patent was not opposed. The Soviet Union succeeded in keeping secret for approximately 30 years the existence and industrial implementation of the most advanced and economically expedient method.

The main bodies engaged in developing the centrifugal method in the Soviet Union were the Central Design Bureau of Machine Building Industry, the Experimental Design Bureau of the Gorky Motor Vehicle Plant, the All-Union Institute of Aircraft Materials, Kurchatov Institute of Atomic Energy, and Center No. 813 which is now called the Ural Electrochemical Center. The first industrial plant in the world, equipped with gas centrifuges, was built and commissioned in 1960-1964 at Sverdlovsk-44 (Verkh-Neivinskiy Settlement). First pilot centrifugal plants appeared in other countries only 10 years later. During 30 years all of the four separating facilities were brought over to the gas centrifugal technology. The technological renovation was effected, mainly, by shutting down and removing the gas-diffusion equipment and installing gas centrifuges on the same floor areas. Concurrently, the first models of gas centrifuges, which had served their life time, were replaced with centrifuges of better performance [94].

The transition to the gas centrifugal technology, carried out in 1966-1972, made it possible to increase the separating capacity of the enterprises by 2.4 times and reduce the overall consumption of electric power by 8.2 times. New building activities were minimized. Five generations of gas centrifuges were developed during that period. In the latest model the specific consumption of electric power per SWU was 25 times smaller compared with gas-diffusion machines. Due to the efforts of the designers and production engineers, the main equipment of the centrifugal technology did not require service and repair jobs throughout the prescribed service life of the machines, namely ≥ 15-20 years.

The exploitation of the gas-diffusion plants ceased in 1992. At present the separation capacities of the enterprises producing enriched ^{235}U using the gas centrifugal technology are distributed as follows, in percent [94]:

Ural Electrochemical Center 40
Siberian Chemical Center 14
Angarsk Electrolysis Center 8
Electrolysis Plant at Krasnoyarsk-45 29

In 1988 a decision was made to stop completely the production of highly enriched uranium for military purposes.

The electromagnetic method did not find industrial application for producing enriched uranium but is widely used for separating stable and radioactive isotopes. This method is based on the laws of the motions of charged particles in an electric and a magnetic field. It employs the same principle of mass separation, as any mass spectrometer. Appreciable quantities of isotopes are produced on large separating setups operating on the principle of Dempster's mass spectrometer[30], with the focusing and deflection of the ion beam through 180° by a transverse magnetic field (Fig. 14). In small setups the ion beam is turned through 60 and 90°. The main advantages of the electromagnetic method of separating isotopes over the molecular-kinetic methods are as follows:

- high separation factor in one cycle;
- possibility of the simultaneous enrichment of all isotopes of the element being separated in one separating setup;
- versatility: one and the same setup can be used to separate stable and radioactive isotopes of a great variety of elements;
- independence of individual separating setups;
- possibility of separating very small quantities (milligrams) of substances.

The disadvantages of the electromagnetic method are:

- low efficiency;
- small substance utilization factor in one separation cycle;
- appreciable irrevocable losses of substances;
- relatively high power and exploitation costs.

With the help of the electromagnetic method appreciable quantities of ^{235}U were produced in the USA already during the war. A large plant was built in Oak Ridge, consisting of several independent setups for magnetic separation of uranium. Later, these setups were used for separating isotopes of other elements. Magnetic field intensity had to be changed severalfold in one and the same setup with an approximately constant path of the ions, i.e., having a fixed position of the ion source and receiver (see Fig. 14), to be used for separating isotopes of heavy and light elements.[31] In the Soviet Union, similar setups were used at Sverdlovsk-45 (*Lesnoi* town)[32] for separating small quantities of uranium isotopes and isotopes of other elements. At present the electromagnetic method of separating various isotopes is widely used in Russia mainly for producing stable isotopes.

The technological processes and physical principles of producing 200 isotopes of 44 chemical elements are discussed in the book *Electromagnetic Separation of Isotopes and Isotopic Analysis* written by N.A. Koshcheev and V.A. Dergachev [96]. The design of the separator (SU-20 setup) and its main components are described. The authors note: " ...adequate resolution of ionic beams for ions with a medium and heavy mass at large divergence angles cannot be ensured in homogeneous magnetic fields. This calls into question the

History of the Soviet Atomic Industry

applicability of the electromagnetic method for producing isotopes on an industrial scale."

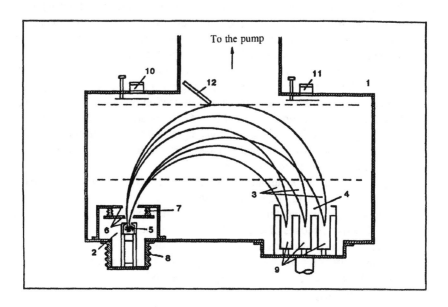

Fig. 14 Schematic of a setup for separating isotopes by the electromagnetic method: 1 – vacuum chamber; 2 – ion source; 3 – ion beams of separated isotopes; 4 – ion receiver; 5 – ion source discharge chamber; 6 – electrodes of the ionic-optical system of the source; 7, 8 – insulators; 9 – receiving pockets; 10, 11 – inspection ports; 12 – diaphragm to protect the pump against contamination with particles of the working substance. The vacuum chamber is evacuated to 10^{-5} mm Hg by powerful diffusion pumps with a total evacuation rate of 20,000 hp.

Strict constraints are imposed upon the receiver design and the extraction of ions from collect pockets. The specific technology of extracting ions from the pockets calls for an individual approach and minute manual work which poorly lends itself to automation. The methods of analytical control are also complicated.

A large SU-20 setup for the electromagnetic separation of uranium isotopes was built and put into operation at Sverdlovsk-45 almost simultaneously with commissioning the first gas-diffusion Plant, D-1, in the Verkh-Neivinskiy Settlement. The main equipment was manufactured at the Leningrad Plant *Elektrosila*. The mass of special magnets was about 6 thousand tons. The magnets were mounted vertically, and arranged in a specially erected many-storied building. Five stories of this building accommodated 20 chambers (4 chambers on each story), and the process of separating uranium isotopes was effected in every chamber. The Institute of Vacuum Engineering developed all necessary equipment to provide a high vacuum (10^{-6} mm Hg) in the chambers. Reliable operation of the chambers was ensured by their rigid structure and the material from which they were manufactured. This material was brass, the wall thickness of the chambers was about 80 mm. Because of the large size of the chambers, about one thousand tons of brass alone were used at the SU-20 setup. Regular operation of the setup required much electric power. Isotopes of many chemical elements from the beginning and the end of the Periodic Table of elements have long been produced on this setup.[33] At the same time, the younger centrifugation method was successfully used for the separation of stable isotopes as well [95, 96].

The centrifugal separation of isotopes is a versatile technique. It is able to separate of both light (B, N, C, O, Kr, Cl) and heavy elements. One of its main advantages is the large scale of isotope production. Without exaggeration, it opens a new era of isotope applications in engineering, science, and medicine. Centrifugal separation leaves the electromagnetic method far behind in efficiency (Fig. 15). In terms of the isotopes presented in Fig. 15, the efficiency of the centrifugal cascade is tens of times higher than that of the electromagnetic setup. By now, Ge, Cr, Zn, and W isotopes have been produced in amounts of more than tens of kilograms. Though relatively young, the centrifugal method yields the main quantity of isotopic products. And, certainly, the lower cost of the isotopes produced in centrifuges should be pointed out. Nevertheless, the uranium centrifuge designed for operation on UF_6 cannot be used effectively for separating stable isotopes of other elements. Centrifuges with individual properties were designed for the above isotopes. Moreover, the substance filling the centrifuge must not cause corrosion of the construction materials. The range of these substances is restricted.

Main Consumers of Enriched ^{235}U

The first uranium bomb exploded at the Semipalatinsk test site on October 18, 1951, was made from a few kilograms of 90%-enriched ^{235}U, produced at the first diffusion plant in 1949–1950 [84]. This happened 2 years after the detonation of the first Soviet plutonium bomb. From the very start the production of the Center was oriented to one consumer of highly enriched uranium – the Ministry of Defense [97]. Therefore, after producing highly enriched uranium from UF_6, metallic ^{235}U was produced and used, like plutonium to make charge components for atomic bombs. Since the beginning of 1951 the gas-diffusion plant could produce approximately 1 kg of highly enriched uranium per day [84]. By that time at other PGU enterprises (at Plant No. 12 and NII-9) a technology was developed for producing items from metallic uranium to assemble uranium blocks for charging the first graphite-moderated uranium pile which operated since 1948 at Chelyabinsk-40 (Center No. 817). The properties of metallic uranium and plutonium were studied, the technology was developed, and the appropriate production facilities were commissioned at Plant V of Center No. 817 for making components of atomic bombs both from metallic plutonium and from metallic uranium, after its 90%-enrichment in ^{235}U.

The uranium bomb was made following the already known technology for a plutonium bomb. The ^{235}U-based atomic weapon was created through the combined efforts of Centers Nos 813 and 817, Plant No. 12, Design Bureau-11 and various plants as well as of the LIPAN, NII-9, NII-10, and the plants producing UF_6. It had to be taken into account that the minimum critical mass of highly enriched metallic uranium (density 18.8 g/cm^3) was substantially higher than that of metallic plutonium with a density of 19.6 g/cm^3.[34]

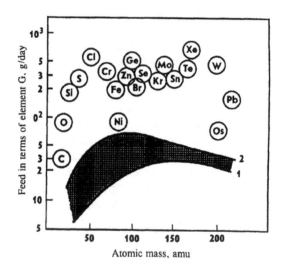

Fig. 15 **Comparative efficiency of the industrial SU-20 setup (1) and the centrifugal cascade-240 (2) in terms of the feed flow.**

Therefore, the uranium nuclear charge was larger than the plutonium one. However, since the early 1950s, new ^{235}U users, other than the Ministry of Defense, materialized, contrary to the case with ^{239}Pu. For instance, Laboratory V, on the basis of which the FEI Institute was founded, was given an assignment, before Plant D-1 was started, to develop a nuclear reactor on weakly enriched ^{235}U for the first atomic power station in the USSR [98, 99].[35]

Already in 1949 decisions were taken to develop research and pilot production reactors, using enriched ^{235}U in the fuel elements loaded into the reactor core. In these reactors, thermal neutron flux density must be higher than in the first production reactor of Center No. 817, which worked on natural uranium. Reactors of this kind working on uranium with $\geq 2\%$ enrichment functioned already in 1951–1952.

The production of tritium in nuclear reactors (the main isotope required for creating an effective thermonuclear weapon) demanded enriched uranium too. A reactor of this kind (AI at Center No. 817) was put into operation on December 22, 1951, with the use of 2%-enrichment uranium blocks as nuclear fuel [98]. As early as 1949 Kurchatov wrote: "For successful designing of high-intensity reactors it is necessary to solve a very complicated engineering problem: to create fuel elements for the physical and technical research reactor" [98]. In this thermal-neutron physical reactor, which since April 1952 became the main nuclear-physical experimental facility at the LIPAN, fuel elements made of weakly enriched uranium had a complicated structure. Their shape and size changed under the effect of neutron irradiation in the reactor. Thanks to the efforts of scientists, special dispersed fuel elements were produced by hot press molding of uranium and magnesium powder[36], which did not change their size in reactors.

A system of research reactors was created not only at the LIPAN, but also in other research centers of Russia, the Ukraine, Byelorussia, and other Republics. In 1955–1958, on the initiative of Kurchatov, decisions were made to build 20 research nuclear reactors in the Soviet Union and elsewhere. All of them were intended to work on enriched ^{235}U. Atomic ice breakers and nuclear reactors for the navy used enriched uranium as well. Experts note [101] that Soviet submariners reached the North Pole many times, and a group of submarines had accomplished a historically unprecedented round-the-world voyage without surfacing.

Since the late 1950s, atomic power stations became numerous consumers of enriched ^{235}U. Nuclear fuel of enriched uranium is supplied to atomic power stations in some foreign countries. In the republics of the former Soviet Union the proportion of electric power generated at atomic power stations was, of January 1, 1993, 78.2% in Lithuania, 29.4% in Ukraine, and 11.8% in Russia.

The efforts mounted in the post-war period to produce enriched ^{235}U for nuclear weapons allowed the Soviet Union not only to live in peace, but also to create an industry for producing nuclear fuel to meet the needs of the national economy. The decision taken in 1988 to stop the production of highly enriched uranium released excess potentials of the plants dealing with the separation of uranium isotopes. They make it possible to export enriched uranium and fully satisfy the needs of the Russian nuclear power engineering (in 2005).

The tremendous efforts made by all those who participated in the production of enriched uranium were not in vain.

NOTES

1 ^{239}Pu is produced by the β-decay of ^{239}Np.

2 From 1946 to 1949 deputies of the Head of 9th Directorate were A.D. Zverev and A.I. Leipunskiy.

3 Worked in the USSR from 1945 to 1955; later – President of the Academy of Sciences of the GDR and a Foreign Member of the USSR Academy of Sciences.

4 The former Chief Designer of the Design Office of Hydraulic Machinery of the Leningrad Metallurgical Works, Head of a Chair of the Polytechnical Institute; in early 1946 he headed the Leningrad Division of Laboratory No. 2.

5 Most prominent Soviet mathematician, Deputy Manager of Laboratory No. 2.

6 They worked in the USSR from 1945 to 1955 and were awarded State Prizes; M. Ardenne was awarded twice (in 1947 and 1953) [5].

7 To settle all issues related to supplies, Institutes A and G were included into the Sinop and Agudzery facilities managed by S.A. Topolin and S.M. Zhdanov.

8 A.I. Churin graduated from the Leningrad Electrical Engineering Institute named after V.I. Ulyanov (Lenin), he headed the Uralenergo System.

9 Deputy Chairman of the Council of Ministers in Stalin's government.

10 Head of the Design Office of the Leningrad plant "Elektrosila"; he was appointed the Chief Designer of electromagnetic separation plants.

11 Ionic sources were developed at the Leningrad Physicotechnical Institute (by A.F. Ioffe, V.P. Zhuze, V.M. Tuchkevich, Yu.A. Dunaev, and others).

12 On their basis the Sukhumi Physical Engineering Institute was organized.

13 Director of the Moscow Plant for Hard Alloys, S.P. Soloviev; Chief Engineer, G.N. Levin. The LIPAN researchers were I.K. Kikoin, V.S. Obukhov, V.Kh. Volkov, and others.

14 In 1944–1949 he worked at the Institute of Mechanics; from 1949 he was the Head of Department; from 1960, Deputy Director of the Laboratory of the LIPAN; from 1962 he was Academician, Vice-President of the USSR Academy of Sciences.

15 This Plant was built in 1938 near the mouth of the Cheptsa River (tributary to the Vyatka River); it was called *Kirchepkhimstroi*. It was transferred to the Ministry of the Medium Machine Building Industry from the Ministry of Chemical Industry in 1958.

16 In the eighteenth century Demidov, a well known Russian factory-owner, built a metallurgical plant at the Verkh-Neivinskiy Settlement, which operated on the power of the Neiva River blocked with a wide earth dam in an area between two hills.

17 The Leningrad Compressor Plant developed a 24-stage, and the Gorky

Machine Building Plant developed a 30-stage diffusion machine.

18 Mounting 24 centrifugal wheels on one shaft and a large number of (curvilinear) ducts for the supply and discharge of gas in a single housing of the machine, as well as of refrigerators and flat porous partitions assemblies behind each compressor, added much to the complexity of the system and lowered its efficiency. Two separating machines manufactured at the Leningrad Compressor Plant proved to be inoperative.

19 Before 1954 all machines and even nuclear reactors developed at the Gorky Machine Building Plant had the index LB (Lavrentiy Beriya); later the index was changed for OK (Russian abbreviation for *Osobaya Konstruktsiya*, i.e., "Special Design"): OK-7 machines, OK-180 nuclear reactors, etc.

20 During the war he was the Chief Engineer at the Uralmash, and after the war he was appointed Director of the Leningrad Compressor Plant.

21 On request of A.S. Elyan, Director of the Gorky Machine Building Plant, the appointment was invalidated in view of troubles with starting OK-7 cascades, and P.S. Mikulovich was transferred from the Uralmash to become the Head of the Shop. From 1951 till 1969 I.I. Afrikantov was a Chief Designer at the Experimental Design Bureau of the Gorky Machine Building Plant.

22 He worked first in Sukhumi, and later was Director of the State Union Research Instrument-Making Institute (SNIIP) in Moscow.

23 Prominent engineer of the People's Commissariat of Nonferrous Metals. Till 1947 he was Chief Engineer at *Glavnikelolovo* and *Glavnikelkobalt*; in 1947–1958 he was Chief Engineer and Deputy Head of the Second Directorate of the PGU; from 1953 till 1956 he was Chief Engineer of the Fourth Chief Directorate of the Ministry of Medium Machine Building Industry (production of ^{235}U and plutonium). Later he was Scientific Secretary of the Scientific and Technical Council.

24 Later on M.P. Rodionov was Director of Center No. 813, and then Director of the Institute of Physics and Power Engineering in Obninsk. When he left, I.D. Morokov was appointed Director of Center No. 813. When he left to become Deputy Minister of the Medium Machine Building Industry, the duties of Director of Center No. 813 were performed during 26 years by A.I. Savchuk.

25 In late 1949 Plant D-1 was reorganized into Center No. 813 which also included Plants D-3 and D-4, and other facilities.

26 Plant D-1 functioned till the end of 1955, when it was stopped in connection with an increase of the volume of production and with starting-up more powerful diffusion plants.

27 The serial number (numerals) denoted the abbreviated year after the war.

28 The final stage of the "Dropshot" Plan stipulated occupation of the USSR and other Socialist countries in Europe by the invasion of 164 NATO divisions, including 69 American divisions, from west and south.

29 Later the Experimental Design Bureau of the Leningrad Compressor Plant was called Central Design Bureau of Machine Building Industry, and the Experimental Design Bureau of the Gorky Machine Building Plant was first called Experimental Design Bureau of the Gorky Motor Vehicle Plant, and then Experimental Design Bureau of Machine Building Industry.

30 A.J. Dempster was a Canadian-born American physicist who built the first mass spectrometer in 1918. He discovered a number of isotopes of potassium, lithium, and other elements. In 1935 he discovered ^{235}U.

31 The mass of the particles collected in the isotope receiver is proportional to the squared magnetic field intensity.

32 The decision of the Soviet government to build Plant No. 418 for the electromagnetic separation of uranium isotopes was taken in late 1947. The first Director of the Plant was D.E. Vasiliev. The chief designer of the Plant was GSP-11. Special equipment was developed under the guidance of D.V. Efremov, Deputy Minister of the Electrical Engineering Industry. The vacuum equipment was developed and designed at the Institute of Vacuum Engineering under the guidance of S.A. Vekshinskiy. The general research management was effected by the Department of the LIPAN headed by L.A. Artsimovich.

33 Isotopes of various chemical elements are separated using the SU-20 setup for creating the State Stock of Isotopes.

34 For a sphere from metallic uranium with 93.5% enrichment, surrounded with a metallic uranium reflector, the minimum critical mass is about 16.5 kg instead of 6 kg for metallic plutonium [100].

35 It was commissioned on June 27, 1954 (the thermal power of the reactor was 30 MW).

36 This technology was developed at Plant No. 12 under the guidance of R.S. Ambartsumyan and A.M Glukhov, scientists from the All-Union Institute of Aircraft Materials (VIAM).

CHAPTER 9

First Soviet Heavy-Water Nuclear Reactors

Frédéric Joliot-Curie, a prominent French scientist, was the first to see the possibility of creating a natural uranium-fueled nuclear reactor moderated by heavy water. He carried out experiments on uranium with a relatively large quantity of heavy water. At that time the world reserve of D_2O was 180 kg, and all of it was in France, where F. Joliot Curie, Hans von Halban, and Lew Kowarski started investigations on building a heavy water reactor in 1939-1940.

As I.V. Kurchatov wrote in 1943 [102], in the Soviet Union it was taken for granted that a chain nuclear reaction cannot be realized in the system uranium-D_2O. It was believed that uranium enriched with ^{235}U was prerequisite for the reactor operation. This conclusion was based on the calculations of Professors Yu.B. Khariton and Ya.B. Zeldovich who demonstrated that for the chain reaction to develop in a uranium–heavy water mixture, it was necessary that the cross-section of thermal-neutron capture by deuterium nuclei should be $\leq 3 \times 10^{-27}$ cm^2. Soviet physicists could not measure the neutron-capture cross-section: there was neither a technical basis nor a sufficient quantity of heavy water. Therefore, use was made of the experimental values of this quantity, obtained by American scientists. Their measurements showed that neutron capture cross-section was more than three times higher than the threshold value calculated by Yu.B. Khariton and Ya.B. Zeldovich. Therefore, the scientists came to the conclusion that the reaction was not feasible in the natural uranium–heavy water mixture [102]. "The conclusion of our scientists, – wrote Kurchatov, – proved to be erroneous." Analyzing the information on the investigations carried out in Great Britain and the USA on the atomic program, obtained by the Soviet intelligence service, Kurchatov came to a conclusion different from that made by Khariton and Zeldovich. In his report to the Council of People's Commissars he stated [102]: "Halban and Kowarski came to the opposite conclusion, which, certainly, is of tremendous fundamental importance. Halban and Kowarski accomplished, according to the available data, the formation of 1.05–1.06 secondary neutron per a primary neutron, i:e., they realized the conditions of a developing chain process."

The French scientists continued their experiments in Great Britain in 1940. In France, already in 1939, having a sufficient reserve of heavy water[1], they determined experimentally the cross-sections of thermal-neutron capture by deuterium nuclei, which could not be determined in the Soviet Union.

In 1943, Laboratory No. 2 just started a research, and, therefore, the data furnished by the Soviet intelligence service were of great value. Kurchatov personally analyzed them in detail on the instructions of M.G. Pervukhin, Deputy Chairman of the Council of People's Commissars of the USSR.[2] The intensity of work along these lines can be judged from the fact that Kurchatov's report dated July 3, 1943 (Document No. 6 from the Files of the External Intelligence Service of Russia [102]) was based on the analysis of 237 American publications, including:

- 29 works devoted to the separation of uranium isotopes by the diffusion method, which Kurchatov estimated to be the leading one in the USA;
- 18 works on centrifugal isotope separation methods;
- 32 works related to the problem of a uranium–heavy water mixture;
- 29 works on the uranium graphite-moderated reactor;
- 55 materials on the chemistry of uranium, containing data on producing metallic uranium from its oxide, on producing UF_6, convenient for separating uranium isotopes by the diffusion method, on organometal uranium compounds, etc.;
- 10 works on making a bomb from ^{235}U;
- 14 documents on the problems of plutonium and neptunium.

The documents were received through the intelligence service and other channels. They were analyzed, and recommendations were given on the problems of particular interest to the Soviet Union concerning the creation of nuclear weapons [103].

Investigations on the creation of a natural uranium heavy-water reactor were stopped in connection with the German occupation of France. The documentation and the reserve of heavy water were transferred to Great Britain. French physicists, H. Halban and L. Kowarski, demonstrated in Cambridge the possibility of attaining a chain reaction using uranium and heavy water as moderator. Taking into account the investigations of other scientists, a committee headed by G.P. Tomson was established in 1940 in Great Britain for managing an atomic project.

During the war, German scientists under the guidance of K. Heisenberg also worked on the development of a nuclear reactor and made desperate attempts to bring heavy water from Norway [104]. In Norway, in 1934, the Norsk-Hydro

Company was the first in the world to start producing heavy water on an industrial scale, using an electrolysis technology developed by two chemists: Professor L. Trønstad and engineer I. Brun. Deliveries of heavy water to Germany started after the occupation of Norway. In late 1940 Norsk-Hydro received an order from the German Concern AG Farbenindustrie. Till the end of 1941 the delivery order for D_2O was 1000 kg; in 1942 it was 1500 kg. Nevertheless, the Germans could not get a D_2O quantity, necessary for creating a nuclear reactor, because the works producing heavy water in Norway was blown up by demolition groups specially trained in Great Britain. By November 1941 Germany obtained only 500 kg of heavy water. The Germans restored the works, and by 1944 about 15 tons of D_2O were produced, but there was no time to deliver it to Germany [104]. After the war, an experimental heavy-water reactor was started in France in 1948 under the guidance of F. Joliot-Curie. However, the first research reactor on naturally enriched metallic uranium with heavy water as moderator was started in the USA in 1944 [105]. The critical mass was 3 tons of metallic uranium, and the quantity of heavy water was 6.5 tons. The thermal power of the first heavy-water reactor was 300 kW, the mean thermal-neutron flux density was 5×10^{11}

ABRAM ISAKOVICH ALIKHANOV,
Academician from 1943.
From 1945, Scientific Secretary of the Technical Council of Special Committee charged with creation of nuclear weapons. Founder of the Institute for Theoretical and Experimental Physics – ITEF (1945); Director of ITEF till 1968. Scientific Coordinator of the USSR project for creating heavy-water reactors.

$s^{-1} \cdot cm^{-2}$. Comparison of these data with those obtained in the first uranium graphite-moderated pile started in Chicago (46 tons), shows that the heavy-water experimental reactor used an approximately 15 times smaller quantity of uranium. The quantity of heavy water, required for realizing a chain reaction, was more than 60 times smaller than the quantity of extra-purity graphite. Any opportunity to save nuclear materials at that time was very important for the Soviet Union. The Soviet Union had no natural uranium of its own yet, to say nothing of the extremely complicated technology of its enrichment.

The data on the US reactors fueled with natural uranium led to the revaluation of priority investigations at Laboratory No. 2. Along with the continuation of research into enriching ^{235}U and making an atomic bomb from it, a task was set to build a reactor operating on natural uranium for producing

ANATOLIY PETROVICH ALEKSANDROV,
Academician from 1953.
President of the USSR Academy of Sciences from 1975 to 1986.
Director of the Institute for Physics Problems from 1946 to 1955.
Director of Kurchatov Institute from 1960 to 1989. Scientific coordinator of the atomic power industry till 1994.

a new fissile material, plutonium. To realize controlled chain nuclear reaction in reactors on natural uranium, graphite or heavy water should be used as moderator.

In 1947 the second research heavy-water reactor was built in the USA [105]. The charge of naturally enriched metallic uranium was 10 tons, and the quantity of D_2O in the core was 17 tons. The power of this more advanced reactor was equal to 30 MW, and the mean neutron flux was 2×10^{13} s$^{-1} \cdot$ cm^{-2}.

In the Soviet Union heavy-water reactors were developed and built under the guidance of A.I. Alikhanov. Similarly to I.V. Kurchatov, he worked before the war at the Leningrad Physical Engineering Institute, and from August 14, 1943, together with I.V. Kurchatov, V.P. Dzhelepov, L.M. Nemenov, G.N. Flerov, M.S. Kozodaev, and other scientists, Alikhanov was transferred to Laboratory No. 2.

In 1943–1945, three reactor philosophies were investigated at Laboratory No. 2:

- I.V. Kurchatov and I.S. Panasyuk headed research into the creation of a uranium graphite-moderated reactor (water-graphite reactor);
- G.N. Flerov and V.A. Davidenko studied the possibility of realizing a chain nuclear reaction in uranium-fueled reactors with ordinary water (water-water reactor);
- A.I. Alikhanov and S.Ya. Nikitin carried out investigations into a heavy-water reactor.

Reports of Kurchatov, Flerov, and Alikhanov on the three types of nuclear reactors were heard in September 1945, at the first session of the Technical Council of the Special Committee. Uranium-graphite and heavy-water reactors were recognized as promising. The main materials for these types of reactors, apart from naturally enriched metallic uranium, were nuclear-purity graphite and heavy water. The natural ratio of heavy and light water is 1:6,800. The separation of heavy water from natural water is a very difficult engineering problem. The problem of industrial graphite production of graphite was tackled at

Laboratory No. 2 by V.V. Goncharov and N.F. Pravdyuk. Physicochemical investigations for the development of commercial methods of producing D_2O were carried out by A.I. Alikhanov, R.L. Serdyuk, and D.M. Samoilovich. In addition, a special subdivision, Sector No. 4, was established at Laboratory No. 2, where different technologies of D_2O production were developed. The Sector was headed by M.I. Kornfeld. Laboratories for developing D_2O production methods were established at the Institutes of the People's Commissariat of Chemical Industry and of the Academy of Sciences. NII-9, where German scientists M. Volmer and R. Doppel were engaged, was also attached to solving the problem.

Establishing of Laboratory No. 3 of the USSR Academy of Sciences

Laboratory No. 3 was established on the initiative of the Technical Council of the Special Committee. At that time the members of the Council were B.L. Vannikov, A.I. Alikhanov, P.L. Kapitsa, I.V. Kurchatov, V.A. Makhnev (Secretary of the Special Committee), Yu.B. Khariton, V.G. Khlopin, and other scientists. The resolution of the Technical Council, dated October 8, 1945, signed by the Chairman of the Council B.L. Vannikov and by Scientific Secretary A.I. Alikhanov, stated: "Consider it necessary to organize Laboratory No. 3 of the USSR Academy of Sciences under the guidance of A.I. Alikhanov, with the following tasks assigned to it:
- physical research, designing, and building of a uranium–heavy water nuclear pile;
- physical research into thorium–water and thorium–plutonium–water systems for producing ^{233}U;
- physical investigations of β-radioactivity;
- physical investigations of high-energy nuclear particles and cosmic rays.

Submit a draft resolution of the Council of People's Commissars of the USSR, moved by Comrades Meshik P.Ya. and Alikhanov A.I. on the above issue, for approval by the Special Committee."

Concurrently, on the proposal of the commission consisting of M.G. Pervukhin (Chairman), A.G. Kasatkin, N.A. Borisov, A.I. Alikhanov, M.I. Kornfeld, V.A. Kargin[3], and L.S. Genin, the Technical Council entrusted the development of conceptual designs for building commercial facilities for producing heavy water using the following methods:
- distillation;
- isotopic exchange with hydrogen sulfide in combination with distillation;
- isotopic two-temperature exchange between water vapor, water, and hydrogen;
- isotopic exchange with hydrogen sulfide.

The development of the design of a plant for isotopic exchange with hydrogen sulfide in combination with distillation was assigned to the All-Union

Research Institute-42 of the People's Commissariat of Chemical Industry, which had proposed this method. The development of conceptual designs of plants based on the other methods was entrusted to a specially formed group of researchers headed by M.I. Kornfeld.

A decree of the government, dated December 1, 1945, sanctioned the organization of Laboratory No. 3. A.I. Alikhanov was appointed Chief of the Laboratory, and V.V. Vladimirskiy was appointed Deputy Chief (since 1946). To date V.V. Vladimirskiy works at the Institute of Theoretical and Experimental Physics. Some of the researchers from Laboratory No. 2 and other institutes was transferred to Laboratory No. 3. After the establishing of NII-9 and Laboratory No. 2, Laboratory No. 3 became a third research organization dealing solely with the PGU problems.

Before Laboratory No. 3 was established, the State Defense Committee in its decision of September 4, 1945, charged the People's Commissariat of Chemical Industry to organize production of heavy water, or product No. 180, as it was called at that time for security reasons. In accordance with a joint order issued by the PGU and the People's Commissariat of Chemical Industry, the designing and building of an installation and a shop for producing D_2O were started.

An area of about 100 ha was allotted for Laboratory No. 3 in Moscow near Cheremushki, in the country estate of the Menshikov family. In 1947, the staff of Laboratory No. 3 was 300 people. By a PGU order of March 26, 1946, 2 million roubles were allocated in the Title List approved by the government for building Laboratory No. 3 in 1946. Without delay, under the guidance of A.I. Alikhanov, technical assignments were prepared for developing an experimental reactor with the use of heavy water as a neutron moderator. The above-mentioned government decree of March 18, 1946, allocated extra payment for the good work of the builders, workers, technicians, and engineers, distinguished in developing a heavy water production technology, in building the shops, and in manufacturing the equipment. It was ruled that government awards should be given to the workers, engineers, and scientists for their achievements. This exceptional attention to the development of industries for producing heavy water and building the corresponding nuclear reactors can be explained by the fact that heavy-water reactors require 10–15 times smaller charges of uranium. The mining of natural uranium in the Soviet Union was at the stage of organization, and its resources were poorly known. For charging experimental reactor F-1 (Laboratory No. 2) and partially commercial reactor A (Center No. 817), use was made of the ore transported in 1945 from Germany [106].

Creation of an experimental basis was started at Laboratory No. 3. In 1948, the main building was constructed and a cyclotron, a deuteron accelerator with a maximum energy of 12 MeV was started. Intense experimental and theoretical research was launched to ensure the building of a research and a commercial heavy-water reactor in the Soviet Union.

From the very start, the main lines of research into heavy-water reactors at Laboratory No. 3 were as follows:
-   nuclear reactor theory;
-   experimental determination of physical constants necessary for the reactor design;
-   physical and heat-engineering experiments;
-   nuclear fuel burnup and the accumulation of neptunium, plutonium isotopes, and transplutonium elements in nuclear fuel;
-   optimization of the physical characteristics of the reactor;
-   problems of the reliable regulation and control of the reactor power.

Particular attention was given to investigating the radiation stability of heavy water and construction materials. The top-priority task of the laboratory was to provide an experimental basis for carrying out neutron physical experiments.

A.P. Rudik, a veteran of Laboratory No. 3, a prominent specialist in designing thermal-neutron reactors, noted that a reactor theory was developed under the guidance of I.Ya. Pomeranchuk, engaged at Laboratory No. 2 from 1943 to 1945.

From the very beginning, V.B. Berestetskiy and A.D. Galanin, well-known theoretical physicists, worked under the guidance of I.Ya. Pomeranchuk; Professor L.D. Landau was engaged as a non-salaried member of the staff. To develop a reactor theory, A.I. Alikhanov and I.Ya. Pomeranchuk invited prominent theoretical physicists from other institutes, A.I. Akhiezer, I.I. Gurevich, A.B. Migdal, I.E. Tamm, and others.[4] A.P. Rudik recalls that by the end of the 1940s a consistent nuclear reactor theory was developed. Among the aspects treated were neutron deceleration, resonance neutron absorption, heterogeneous reactor theory, fast breeding, problems of reactor regulation in the case of delayed neutrons, and the like. These problems were treated later in [107] and other publications.

Research Heavy-Water Reactor of Laboratory No. 3

The creation of the first uranium graphite-moderated pile and the development and building of a heavy-water reactor were carried out mainly under the supervision of Section 1 of the PGU Scientific and Technical Council. During 1947, when a heavy-water reactor was designed, Section 1 of the Scientific and Technical Council discussed the reports of its main designers.

On January 16, 1947, A.I. Alikhanov, Head of Laboratory No. 3, and B.M. Sholkovich, Chief Designer of the Experimental Design Office "Gidropress" of the Podolsk Heavy Machinery Works, reported on the design assignment for developing an experimental reactor. The instructions of Section 1 included:
-   to A.I. Alikhanov: make and submit additional physical calculations; to I.V. Kurchatov: consider these calculations and submit them for approval

together with the design assignment;

- to A.I. Alikhanov, and also to A.P. Zavenyagin and P.Ya. Antropov, PGU Deputy Heads: determine the maximum length of uranium rods, feasible for manufacture at Plant No. 12 in Elektrostal;
- to B.M. Sholkovich: consider additionally the problem of the maximum 500 kW power of the experimental reactor.

Among the persons invited to the meetings of Section 1 were not only the designers, but also members of the PGU Administration and Ministers of other industries, whose plants and institutes were to manufacture the equipment and instruments for heavy-water reactor within the shortest possible time.

Unlike the first experimental uranium graphite-moderated reactor F-1, built at Laboratory No. 2, the physical deuterium reactor was a complicated physical system distinguished by numerous specific design features and an appreciable quantity of equipment. A characteristic feature of heavy-water reactors is the formation of a fulminating mixture in them, which has to be removed and burned. These reactors must meet especially stringent requirements in terms of tightness and safety. Another specific feature is that in the vessel of the reactor, or in any other equipment of the reactor loop, a spontaneous chain reaction may occur. The critical mass may induce this reaction, when the quantity of uranium in the $U–D_2O$ system is 1.5-2 tons. Therefore, the designers had to be very careful.

The high cost of heavy water called for the minimization of leakages and losses of this neutron moderator which functioned simultaneously as a heat carrier. Checking of the equipment for tightness (in the first loop) had to be especially strict. Because a commercial reactor was designed and built following the design of the physical deuteron pile, it is reasonable to consider the design and scheme of operation of the reactor built at Laboratory No. 3 in more detail (Figs 16–18).

Heavy water was poured into a thin-walled aluminum cylinder 1,175 cm in diameter, 195 cm high, with the walls and bottom 3-3.2-cm thick. The bottom of this tank was placed on a 100-cm thick graphite stack. The side graphite reflector 2 was also 100-cm thick. The entire graphite stack was placed in an external tight steel housing 3, in which vacuum had to be maintained. The design allowed the replacement of the thin-walled aluminum cylinder. Uranium rods 4 were suspended from rotary upper plate 7, above which lead protection cover 5 with four slit shutters 6 was disposed. Tight stainless steel plate 8 with four slit gaskets for changing the rods was arranged above the lead cover. This design of the plate and shutters made it possible, when the shutters were opened and the rotary plate was turned, to bring any uranium rod under a slit and change it without dismantling the upper protection cover of the reactor. The plate was provided with a large number of additional openings that allowed the operator to change the space of the uranium-rod grid. D_2O was fed to the reactor through delivery pipe 9 from below and drained in the upper part of the

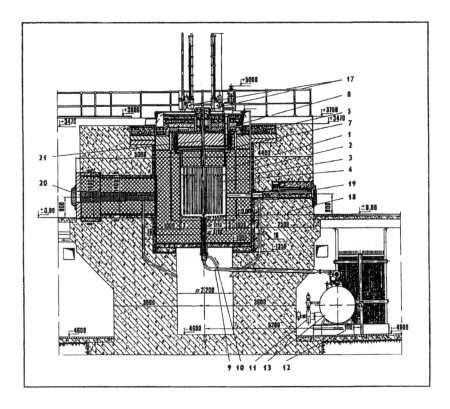

Fig. 16 **Longitudinal section of physical deuterium pile [112] (explanations are given in the text).**

reactor. Draining pipe *10* went down along the reactor axis and then ran inside the delivery pipe. Then the water flowed through circulation pumps *11* to heat exchanger *12* and returned to the reactor along the delivery pipe. Because the rate of forced D_2O circulation in the loop was low, the power of the reactor depended on the possibility of natural convection, the rated power being 500 kW. The rated capacity of the circulation pumps was 25 m^3/h. Leakages of heavy water through the pumps did not exceed 10 g/day. D_2O was stored in standby container *13*. The circulation of helium, placed above the heavy water, was effected by gas purge pumps. Helium entrained the fulminating mixture formed in the reactor together with heavy water vapors which precipitated in condenser *15*, and the fulminating mixture was burned in contact apparatus *16* on a palladium catalyst. Four channels with neutron absorbers in the form of cadmium rods were used for the manual and automatic regulation of the reactor power. Motor drives *17* and regulator sensors were mounted outside the reactor. The thickness of concrete in side protector *18* of the reactor was 2.5 m. Through horizontal channels *19* neutron beams and γ-radiation beams were brought out for research purposes. Three channels were extended through the graphite reflector to the reactor tank. The neutron beam was brought out from the center of the reactor through vertical blind channel *21*, 9 cm in diameter. Horizontal graphite column *20*, 1.4×1.4 m in size, at the center of the side protector was sufficient for carrying out a large number of experiments. The reactor was equipped with various systems to control D_2O and gas flow rate and moderator filling level and measure temperature and pressure. All of the vessels and pipelines were tested for vacuum tightness. Particular attention was given to measuring a fulminating mixture concentration in the gas. The reactor could be drained and dried in the case of repair jobs. Vacuum pumps *22* made it possible to freeze out heavy water vapors in liquid nitrogen traps *23*.

The reactor building was erected and the main equipment mounted during 1948. The reactor was commissioned in April 1949. The physical deuteron pile was used to work out and design the first commercial heavy-water reactor which was to be built in Southern Ural (Center No. 817).

For choosing an optimal scheme of charging, uranium rods of 2.2 and 2.8 cm in diameter with a 0.1-cm thick aluminum jackets were used in the experimental pile. The length of the uranium rods was 162.5 and 160 cm, respectively. The spacing between the uranium rods in D_2O was varied from 6.3 to 16.26 cm. The number of uranium rods constituting a critical volume for different grid spacings varied more than threefold [112]. The minimum critical mass − 86 rods, 28 mm across, with a grid spacing of 16.26 cm, and 107 rods, 22 across, with a grid spacing of 12.7 cm − was ≤1.5 tons of uranium instead of 46 tons in the experimental uranium-fueled graphite-moderated pile.

In the commercial heavy-water reactor, the diameter of uranium blocks was taken to be 22 mm. The experimental data on the size of the reactor core confirmed almost completely the calculations performed at Laboratory No. 3 and reported by A.I. Alikhanov to the PGU Scientific and Technical Council on May 13, 1946. He reported that for building an experimental heavy-water reactor it was necessary to have 4 tons of heavy water, 2.5 tons of metallic uranium items, and 50 tons of graphite, partly serving as a neutron reflector. The experimental reactor was planned to have a vessel about 2 m across and approximately 2.3 m high. With a reflector, these dimensions were to be 3.3 and 4 m, respectively, with a biological shield of 9 and 5 m. Emphasizing the advantages of heavy-water reactors over uranium-graphite ones, Alikhanov wrote in his memorandum to the PGU Scientific and Technical Council: "A commercial system requires different quantities of uranium for different designs of the pile, which generally total ≤10 tons in contrast to a graphite pile, where approximately 150 tons are required. Another advantage of D_2O as a moderator is that the moderator is a liquid, the phase that has long been used in power

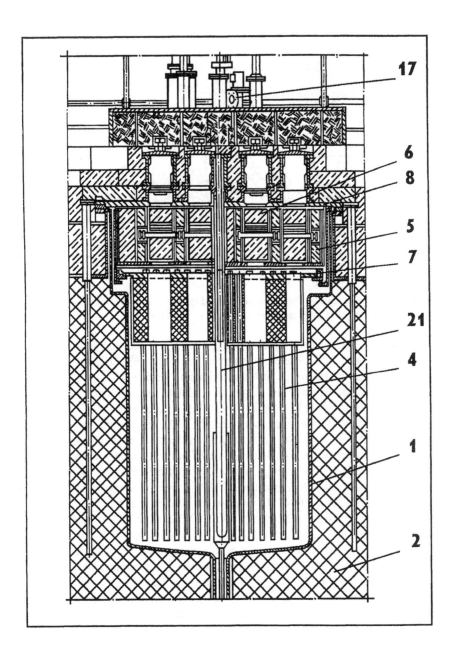

Fig. 17 Sectional view of the reactor, illustrating the arrangement of slit shutters [112] (explanations are given in the text).

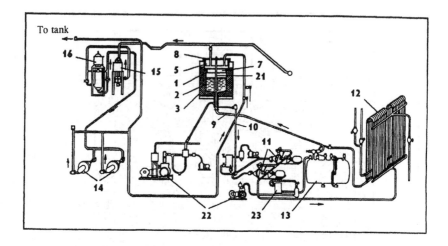

Fig. 18 Scheme illustrating the heat removal system and operation of other systems of the reactor [112] (explanations are given in the text).

engineering." It was contemplated that the leakage of neutrons from the reactor would be about 14%. Therefore, to avoid inefficient ^{235}U burnup, Alikhanov suggested that the reactor be surrounded with a "thorium hood": "20 or 30 tons of thorium oxide or metal will absorb 80–90% of these escaping neutrons and yield approximately 2–3 g of ^{235}U a day. With a reactor power of ca. 30 MW, 25–30 g of plutonium will be produced daily." The reporter pointed out that the power of the reactor had been determined on the basis of rough estimates, and could be revised on the completion of a design for a commercial reactor.

Later, the Heat-Engineering Laboratory experimental reactor was modified. In 1957 its core vessel was modified, ^{235}U enrichment was increased to 2%, and the power raised to 2.5 MW. To increase the physical parameters of the reactor, in 1963 its core started to be charged with 80%-enriched ^{235}U. Uranium blocks were loaded into 64 fuel channels (7 blocks per channel) with a spacing of 13 cm. The quantity of uranium in the reactor core was 3.3 kg and the content of D_2O was 4 tons in the reactor vessel and 5 tons in the entire loop. The D_2O flow rate was 100 m^3/h. The temperature of the heat carrier was 50°C at the input and 70°C at the output. For regulating the power, the reactor was provided with 7 channels of neutron-absorbers. These parameters made it possible to increase the neutron flux density from 2.5×10^{12} to 4×10^{13} s$^{-1} \cdot$ cm^{-2}.

Under the scientific guidance of the Institute of Theoretical and Experimental Physics (ITEF), 10 MW TVR-S heavy-water reactors were designed. With the assistance of the Soviet Union, reactors of this type were built in 1959 in China and Yugoslavia [113].

At the ITEF the reactor became the main experimental basis for research in nuclear physics, solid-state physics, and nuclear reactor physics. The main scientific achievements of the Institute were described in a paper by V.V. Vladimirskiy [113], Deputy Director, Corresponding Member of the Academy. In 1960, ITEF reactor yielded an intense polarized beam of thermal neutrons for the first time in the USSR with the use of magnetized cobalt mirrors as reflectors. In 1964, a phenomenon of space reflection symmetry violation was discovered in the process of radiation capture of neutrons by the nuclei, caused by a weak nucleon–nucleon interaction. This result was confirmed in 1972 by a group of American physicists. The early theoretical study was carried out by Yu.G. Abov and P.A. Krupchitskiy, both from the ITEF. They got Lenin Prizes. This discovery was registered officially. Another discovery was made at the ITEF in 1972: the phenomenon of the P-odd asymmetry of divergence of fission fragments, also caused by a weak nucleon–nucleon interaction, was discovered.

The Institute of Theoretical and Experimental Physics with its experimental reactor was the father of a whole series of investigations that have been carried out in many research centers of the world. One example is an experiment with beams of resonance and ultracold neutrons. Since 1960, neutron choppers with rotors suspended in a magnetic field were developed and tested at the Institute. The rotors were placed in evacuated casings and rotated in synchronism without friction. Using the neutron time-of-flight method, this neutron chopper allowed one to measure the cross-section of the interaction of neutrons with different nuclei depending on the energy of the neutrons. The idea of creating neutron choppers belonged to V.V. Vladimirskiy and was implemented at the Institute by a group of researchers headed by S.M. Kalebin. The four-rotor neutron chopper developed at the Institute was mounted on a unique SM-2 reactor in Dimitrovgrad (Research Institute of Atomic Reactors). The neutron flux density was as high as reached 5×10^{15} s$^{-1} \cdot$ cm^{-2}.

These advances made it possible to have in the Soviet Union a kind of a "factory" of effective neutron cross-sections. In addition to fundamental research, applied investigations were carried out on the ITEF experimental reactor. For a long period of time different radioactive isotopes were produced in the reactor. In 1987, the reactor was shut down. In connection with changes in the topical trend of research, the reactor is expected to be used only as subcritical stand for experiments with charged-particle beams obtained on accelerators.

Heavy-water reactors offered a number of advantages over uranium-graphite reactors. It was of particular importance that they required a substantially smaller charge of uranium and a smaller quantity of the main core material, i.e., neutron moderator per unit power. However, the operation of these reactors required special measures for ensuring nuclear, engineering, and radiation safety, because hydrogen was formed under the action of neutrons, and a fulminant mixture originated. The problems of choosing sites for building these reactors were discussed once and again in 1946 at the PGU Scientific and Technical Council. After numerous discussions it was decided to build an experimental reactor at the site of Laboratory No. 3, and a commercial reactor at Center No. 817 in Chelyabinsk-40.

First Design Institute in Atomic Industry

The Leningrad State Special Design Institute-11 was at that time the sole general designer of all reactors. It is still the general designer of many production facilities in the atomic industry, including radiochemical plants and plants for separation of uranium isotopes. By the order of the People's Commissariat of Heavy Industry, No. 183 of October 21, 1933, [114], a special Design Bureau *Dvigatelstroi* was established (6, *Naberezhnaya Krasnogo Flota*) under the jurisdiction of the Chief Military-Mobilization Directorate of the People's Commissariat of Heavy Industry. In 1938 the Bureau was named State Special Design Institute-11, and by Decree No. 4 of January 21, 1939, it was placed under the People's Commissariat of Ammunition. During the war the Institute was evacuated to Kirov City, where it worked to orders from 24 People's Commissariats, designing 89 military works and complexes. In 1944, the Institute was returned to Leningrad, and by Decision No. 996 of the State Defense Committee of September 4, 1945, it was placed under the PGU. In 1941–1945 the Institute was headed by A.I. Gutov, the Chief Engineer was F.Z. Shiryaev. After the Institute had been subordinated to the PGU, Gutov remained to be its Director, and V.V. Smirnov was appointed a Chief Engineer. The first designs of the Institute were orders to reconstruct some of the oldest Soviet plants (for example, Plant No. 48 in Moscow).

The State Special Design Institute-11 included 18 departments and design offices, as well as all necessary technical services. (After reorganization in 1955, six complex design offices were arranged). Since 1946, the office of Deputy Chief Engineer was held by A.Z. Rotshild, A.V. Karandashev, A.A. Chernyakov, and I.Z. Gelfond; and since 1949, by A.N. Matveyev, who had been connected with the atomic industry for many years.

Already at the beginning of 1946 the Institute started to design first industrial centers and residential areas to be built in the Ural region. Sites for Plutonium Center in South Ural, where, in addition to uranium-graphite reactors, the first heavy-water reactor was built later on, were chosen, in accordance with the decision of the PGU Scientific and Technical Council already on June 13, 1946 (Minutes No. 21). After A.A. Chernyakov presented a design assignment for Plant No. 817, as it was then called, it was decided:

- to provide land for locating the plant facilities over an area of ca. 200 m² on the southern shore of Lake Kyzyl-Tyash;
- to locate the first uranium-graphite reactor A at a height of 270.7 m in the middle of the site between the other facilities;
- to dispose a radiochemical plant at a distance of approximately 2 km from nuclear reactor A;
- to dispose water-treatment sites 1.7 km apart from nuclear reactor A;
- to locate a residential settlement at a distance of 8–9 km from reactor A on the shore of Lake Irtyash.

In the design assignment, provision was made for the sites of settling ponds and fragment storages, i.e., storages of radioactive products of fission, produced in nuclear reactors. In the assignment for power supply, approved by the PGU Scientific and Technical Council, it was envisaged to have a "System of power supply for Plant No. 817 to power two substations of the plant from three independent power transmission lines: Chelyabinsk Heat and Electric Power Plant, Kyshtym and Ural Substations (second priority), and from its own permanently operating Heat and Electric Power Plant with two steam turbines, 2–2.5 kW each." On the same production site of the future Center No. 817, not far from a water-treatment shop, a site was chosen for building a commercial heavy-water reactor OK-180.

More stringent safety requirements were placed on Center No. 817: protection from external effects (crash of an aircraft, etc.), specific requirements to transient conditions (bringing to the rated power, reducing the power and shutting down the reactor). It was necessary to obviate the formation of thermal stresses exceeding the permissible ones in the metallic structures of the reactor, including its vessel. Therefore, besides the above-cited power supply sources of Center No. 817, an additional power supply source was envisaged for reactor OK-180, storage batteries.

In addition to the nuclear centers in the Southern and Central Ural, the later designs developed at State Special Design Institute-11 were used to build the main facilities of Centers No. 816 (Tomsk-7) and No. 815 (Krasnoyarsk-26), and also the first Atomic Power Station in the world at Obninsk, whose 40-year anniversary was celebrated in June 1994. The Siberian, Leningrad, Kursk, Ignalina Atomic Power Stations, as well as a unique distillation plant with fast-neutron reactors (BN–350) in Kazakhstan were built in accordance with the design documents issued by the Institute. V.A. Kurnosov, General Director of the All-Russia Design, Research, and Engineering Association[5], recalls that a dozen of secret and a great many of open towns of the Ministry of Medium Machine Building Industry were designed by the specialists of the Leningrad Design Institute. Examples are Chelyabinsk-40, Sverdlovsk-44 and Sverdlovsk-45, Krasnoyarsk-26, Penza-19, Shevchenko in Kazakhstan, Navoi in Uzbekistan, towns of research centers (Arzamas-16), and academic campuses in Novosibirsk, Obninsk, Sosnovyi Bor, and Dimitrovgrad.

Building of a Commercial Heavy-Water Reactor

The experimental heavy-water reactor, or, as it was then called, physical deuteron pile, was developed simultaneously with a commercial deuteron pile under the scientific guidance of Laboratory No. 3. The participation of leading institutes of the Soviet Union and the plants of the Ministry of Chemical Industry in D_2O production, as well as the already functioning facilities and experimental shops for producing articles from metallic uranium at Plant No. 12, enabled the PGU Scientific and Technical Council to issue in September 1946 recommendations on the terms of commissioning a commercial deuteron pile based on the proposals of A.I. Alikhanov and V.V. Vladimirskiy.

A site for building a commercial heavy-water reactor OK-180 at Center No. 817 was chosen in the fall of 1948. Earth excavations were started in the summer of 1949, before the final approval of the technical design. Dozens of institutes and plants participated in the creation of the reactor. Veterans of Laboratory No. 3 note that 56 organizations, including 16 research institutes, were invited to design the process of manufacturing uranium blocks, canned in aluminum.

In the spring of 1950, a full-scale stand for 38 fuel channels was mounted at the Experimental Design Bureau of the Gorky Machine Building Plant. A full-size model of the reactor vessel and a full-size pressure chamber were built. From this chamber water flowed to the lower part of the tank (housing) of the fuel channels to cool the surface of uranium blocks. Fuel channels were made from aluminum alloy. The rate of flow of the heat carrier (D_2O) in the channels was several meters per second. Moderator (also heavy water) was fed to the annular space at a small rate.

The stand was used to test and improve the method of unloading uranium blocks from the fuel channels. Difficulties arose as soon as the reactor began to operate. It was found that on cutting off the circulation pumps uranium blocks might pile up in the delivery pipe, and the channel might be clogged. When four channels were unloaded, the blocks stuck almost completely and did not get into the receiver. Because the residual energy release of irradiated uranium blocks in heavy-water reactors was 6–10 times higher than that in uranium-graphite reactors, their sticking in the unloading channel with insufficient cooling could lead to overheating and even to melting and burning holes in the walls of the channel, and that was a very serious breakdown. It should be recalled that two versions of cooling uranium blocks were offered in the technical assignment for the design of reactor OK-180.

In the first version the moderator was heavy water and the heat carrier was ordinary water. It was believed that it was much simpler to cool uranium blocks in aluminum pipes using ordinary water and that there would be no problems with their unloading. Moreover much heavy water would be saved. This was the main version; the development of its design assignment was entrusted to F.I. Rylin.

In the second version, heavy water was used as moderator and as heat carrier, so that the demand for it increased. In that case, however, the requirements to the tightness of the annular space were less severe, because heavy water, even in the event of leakage, did not come in contact with ordinary water. The difficulty with the second version was that in unloading uranium blocks for transferring them to cooling ponds with ordinary water, heavy water was lost together with the moist blocks. The person responsible for the development of the second version was N.N. Kondratskiy (State Design Institute-11).

G.N. Karavaev and M.M. Kutakov, both of the State Design Institute-11, jointly with N.N. Kondratskiy, proposed an original design of a hydraulic transportation device, operating on heavy water, for unloading uranium blocks. Owing to this engineering solution, preference was given later to the second version.

ALEKSANDR IVANOVICH GUTOV.
General Designer. Director of the institute involved in designing first facilities of the USSR atomic industry.

The stands at the Experimental Design Bureau of the Gorky Machine Building Plant were used to improve the system of hydraulic separation. It was established that blasting of the transportation pipe in a direction opposite to the falling uranium blocks (unloaded from the fuel channels) led to the transfer of an appreciable quantity of ordinary water to the system of the reactor filled with heavy water. Therefore, blasting with helium had to be rejected. A technology was developed for separating uranium blocks from less heavy empty blocks. To this end, two grids were mounted in the hydraulic separator. Heavier uranium blocks settled down on the lower grid, whereas less heavy empty blocks settled down on the upper grid. 1,800 experiments were carried out on the stand of the Experimental Design Bureau; 225,600 blocks were separated with only 3 uranium blocks erroneously identified as empty.

Some equipment was tested at other institutions:
- at Experimental Design Bureau-12 (S.A. Frankshtein) − mockups of reactor protection control systems. This Design Bureau was responsible for the development of reactor protection control systems;
- at the Nevskiy Machine Building Plant (M.A. Agre) − helium gas pumps. Later, these pumps were manufactured at this Plant;

- at Design Office-12 of *Teplokontrol* Trust (V.K. Dmitriev) — all of the measuring instruments.

On the proposal of A.I. Alikhanov a Commission on Corrosion of Heavy-Water Reactor Equipment was established. Its members were R.S. Ambartsu-myan (All-Union Institute of Aircraft Materials), G.V. Akimov and N.D. Toma-shev (Institute of Chemical Physics), and B.V. Ershler (Laboratory No. 3). Following recommendations of the Commission, it was decided that the reactor vessel and the pressure chamber should not be anodized. The main materials from which they were manufactured were aluminum alloys. Heat exchangers were also made from an aluminum alloy. In addition to the heavy-water loop, the design included a second loop, filled with ordinary water. The requirements to the tightness of heat exchangers were extremely severe. Reactors on D_2O were the first reactors provided with heat exchangers. Their chief designer was B.M. Sholkovich.

Thanks to the efforts of numerous collective bodies responsible for the manufacture and delivery of the equipment, the builders and erectors of Center No. 817 were able to start mounting reactor OK-180 and all of its diverse systems in mid-1950.

Start-up Adjustments and Initial Period of the Reactor Operation

Mounting of the main equipment of the reactor was started in 1950. During 1950–1951, the building site of the heavy-water reactor became a working place for many specialists from the Heat-Engineering Laboratory, the Experimental Design Bureau of the Gorky Machine Building Plant, Experimental Design Office "Gidropress", and State Design Institute-11, as well as designers and manufacturers from other works and design organizations.[6] Chelyabinsk-40 became the place of permanent residence of the scientific supervisor in charge of heavy-water reactors.

Since late 1950, Academician A.I. Alikhanov constantly worked at the site, being responsible, jointly with B.G. Muzrukov, G.V. Mishenkov, and other officials from the Center for the supervision of the mounting and start-up adjustments of the reactor. Veterans of Plant No. 37 (B.V. Gorobets, B.A. Kud-ryavtsev, and others) recall that one could often see Academician Alikhanov in the central reactor room and at other mounting sites. His assistants from the Heat-Engineering Laboratory participated in setting-up various control systems of the reactor and in preparing the reactor to commissioning.

The workers of the Heat-Engineering Laboratory prepared all necessary manuals for servicing the sophisticated systems of the reactor. The service personnel of the reactor attended a course of lectures, including the aspects of the reactor physics and operation safety. The lectures were given by executives from the Heat-Engineering Laboratory V.V. Vladimirskiy, N.A. Burgov, and others, designers from the Experimental Design Bureau of the Gorky Machine Building Plant, State Design Institute-11, Experimental Design Bureau-12, and specialists from other organizations which developed the main equipment. Examinations were administered by a special Commission with the obligatory participation of the Research Manager or his Deputy. Not all persons were allowed to service the reactor.

During the start-up tests and adjustments the first loop was filled with ordinary water, the fuel channels were loaded with uranium and empty blocks. Setting-up jobs were carried out for all the reactor systems, including the systems of unloading blocks from the fuel channel and the unloading loop. The efficiency of thermal compensation of the fuel channel was checked by heating the system to 70°C. Research Managers were on duty in every shift. A.I. Alikhanov, V.V. Vladimirskiy, N.A. Burgov, S.A. Gavrilov, A.V. Zinchenko, S.Ya. Nikitin, and P.A. Petrov were among them. A.A. Tarasov was appointed Head of the Facility.

A special horizontal channel was provided in the reactor for carrying out physical experiments. A unique neutron beam of this channel with a relatively large flight length was actively used by the experts of the Heat-Engineering Laboratory for precision measurements of the effective cross-sections of neutron interactions with various materials. A neutron spectrometer with a mechanical chopper, designed at the Heat-Engineering Laboratory under the guidance of V.V. Vladimirskiy, was installed on the neutron beam. The neutron density in that beam was then the highest attainable in the Soviet Union. At first, supervised by V.V. Vladimirskiy, researchers from the Heat-Engineering Laboratory, N.A. Burgov, V.V. Sokolovskiy, I.A. Radkevich, and others worked with the spectrometer. Later, experiments were carried out by researchers from Laboratory No. 5 of the Center G.M. Drabkin, V.N. Nefedov, L.G. Stepanova, V.I. Orlov, and others. Kurchatov attached particular significance to the investigations with the neutron beam. Visiting the Center, he regularly gathered those who participated in the experiments to discuss the results obtained in measuring the cross-sections of neutron interactions with the nuclei of different elements, and in experiments on nuclear isomerism.

The first serious trouble was discovered in the course of the starting-up and setting-up jobs when running the second loop filled with distillate of ordinary water. Leakages of lake water were detected in the heat exchangers. Vibrations of individual pipes in the heat exchanger at a high flow rate of water disturbed the tightness. The chief designer of the heat exchangers (Experimental Design Office "Gidropress") decided that some of the pipes should be removed. The heat exchanger tightness was tested again with ordinary water. Prior to filling the reactor with heavy water, the loop was washed and dried using two tanks of extra-purity (technical) alcohol. The used alcohol was disposed into the Techa River, near which the reactor OK-180 was built at the Lake Kyzyl-Tyash at a distance of about 200 m from the river.

Experiments on attaining the critical state of the reactor core (onset of a self-sustained chain reaction) were carried out with the fuel channels filled with uranium blocks. Heavy water was poured slowly into the loop, and the neutron flux was thoroughly monitored. The experiments were supervised by Alikhanov, Vladimirskiy, Galanin, and Burgov. The total quantity of heavy water in the loop of the working reactor was 30 tons. Heavy water was poured into the first loop through the reactor cover manually from 10-liter containers. Approximately 15 tons of uranium blocks were loaded into the reactor core. The channels for loading thorium were empty. To deaerate the heavy water, the entire first loop was filled with helium before bringing the reactor to the full power. Power boosting was started on October 17, 1951, and in a few days the reactor was brought to the power of 100 MW.

As mentioned above, the reactor core was charged with fuel elements, namely, with uranium blocks, 22 cm across and 75 mm high, in aluminum jackets. The jacket of the fuel element was diffusion-coupled with the uranium core through a sublayer. The coupling technology was developed by the scientists of the All-Union Institute of Aircraft Materials (Moscow, R.S. Ambartsumyan) and of Research Institute-13 (Leningrad, P.P. Pytlyak). Ambartsumyan suggested that the sublayer should be made from phosphate, whereas Pytlyak proposed a nickel sublayer, 2–3 μm thick. 100 fuel elements of each type were manufactured at Plant No. 12 in Elektrostal, and after testing preference was given to Pytlyak's technology. The procedure for checking the tightness between the jacket and the uranium core was developed by Professor S.Ya. Sokolov (Leningrad Electrical Engineering Institute) and engineer B.S. Krychov (Research Institute-13). Krychov's ultrasonic flaw detector was adopted for checking the products manufactured at Plant No. 12.

Uranium blanks, 70 mm in diameter and 400 mm long, were prepared by casting. At first graphite molds were used, then they were replaced by steel ones. Castings were heated to 900°C in electric furnaces, then rolled down to 40 mm in diameter, and cooled in water. The uranium rods thus produced were cut and pressed at 600–625°C to a diameter of 22.5 mm. After that the uranium blocks were turned down and transferred for jacketing. The burnup of uranium in the reactor was up to 160 MW · day/t, and the accumulation of plutonium was about 180 g/ton. On an average, 200 g/ton uranium fission products were produced.

Emergency situations, similarly to the case with uranium-graphite reactor A, started immediately, during the first month of the reactor operation. In November 1951, when the temperature of water in Lake Kyzyl-Tyash lowered to 0–3°C, heavy water in the heat exchangers froze (its freezing point is 3.8°C). Circulation of D_2O in the reactor stopped. The reactor was immediately scrammed. Nevertheless, because of residual power evolution from the uranium blocks in the absence of circulation, the heavy water in the fuel channels began to boil. One can judge about this emergency and other difficulties which arose in the development of reactor OK-180 from a sufficiently dramatic communication of E.P. Slavskiy, quoted in his conversation with A.P. Aleksandrov and Admiral V.N. Chernavin [116]. This happened on the eve of Alikhanov's departure to Moscow on completion of the setting-up jobs on the reactor. During the farewell party, E.P. Slavskiy, who worked since 1949 as Deputy Head of the PGU and was at that time at the Center, was informed that the circulation of heavy water in the reactor had stopped. He addressed right away to those present at the party: "Well, guys, take the bus and let's go... Alikhanov, effusive as he was, at first did not believe that the reactor had got frozen. He shouts: "It's impossible!" − I: "What's impossible? There's no water circulation!"... Our gates were made from cast iron, automatic and there were two heat exchangers behind them, big cylinders. I say: "Let's go there and see." They: "But there's radioactivity there!" "Never mind." I opened the gates, went in, put my hand on the heat exchanger, and found it to be cold instead of being hot: "Well, come up!" And only then did everybody understand what had happened." To preclude D_2O freezing in the heat exchanger in future, it was decided not to allow the temperature in the loop with ordinary water to drop below 8°C. Actually that was not an accident but an emergency situation, which illustrated how difficult it was to operate heavy-water reactor. Thermal stresses which originated on scramming the reactor, especially on the emergency protection operation or on rapid power boosting sometimes exceeded the permissible ones. Therefore, limitations were introduced on the power boosting and dropping rates. The heat exchangers made from aluminum did not last long. They were soon replaced with new ones, made from stainless steel.

The most serious accident occurred on reactor OK-180 in its unloading system, which had been thoroughly tested on the stands at the Experimental Design Bureau of the Gorky Machine Building Plant. When uranium blocks, containing the usual quantity of produced plutonium, were unloaded from several fuel channels, they stuck in the hydraulic transportation device. The accumulation of a large number of blocks with a high residual liberation of heat led to their overheating and melting, to the burnup of the components of the hydraulic transportation device, and to the failure of the entire unloading system. Urgent measures were taken to remedy this radiation-hazardous emergency, and after that the reactor was again ready for operation. The jobs were supervised personally by E.P. Slavskiy, Head of the 4th Department of the PGU, A.D. Zverev, and the Center Directorate. The technology of unloading the reactor was modified. Uranium blocks held in flared-out fuel channels had to be extracted through the top of the reactor into the central room and then brought to the cooling pond for storage, reloading into containers, and preparing them for delivery to radiochemical processing. As a result, the quantity of produced plutonium was reduced.

The reactor worked in different modes. Initially, when the reactor was charged with natural uranium, mainly plutonium was produced. Then, since

1954, after passing over to charging uranium blocks with a 2% enrichment in ^{235}U, thorium was charged into the core for producing ^{233}U. Some of the channels served for producing other isotopes, e.g., ^{60}Co, ^{32}P, etc. This was done on a limited scale, because the physical characteristics of the reactor working on natural uranium had a small excess secondary neutron-multiplication factor, and the quantity of secondary neutrons was sufficient only for sustaining a chain nuclear reaction. Upon passing over to charging uranium with a higher ^{235}U enrichment, it became possible to produce not only ^{233}U (from ^{232}Th), but also tritium (from ^{6}Li) for thermonuclear weapons.

The OK-180 reactor was shut down in 1965. The OK-180 reactor, located in Building 401 at Plant No. 37 was not the only one. Already in 1953, the designers of the Experimental Design Bureau of the Gorky Machine Building Plant started mounting another heavy-water reactor, OK-190, in Building 401a. It was commissioned on December 27, 1955. That reactor worked for 10 years and was shut down on November 8, 1965.

Later, owing to the efforts of the operating personnel of Plant No. 37 and different services of Center No. 817, with the assistance of the NIKIMT (Director Yu.F. Yurchenko), the reactor was disassembled. The experience gained in disassembling the reactor was published in part in 1990 [117]. After the disassembly of reactor OK-190, a virtually new reactor, OK-190m, worked in Building 401a since April 1966 and was shut down in 1986 [118].

NOTES

1   The reserve of heavy water in the Soviet Union was 2 kg only [102].
2   According to the data of the People's Commissariat for Internal Affairs, the Soviet intelligence service obtained 286 secret documents from England and the USA [174]. Kurchatov mentioned seven research centers and 26 American experts as sources of this information [174].
3   V.A. Kargin (1907–1969), chemist, Academician since 1953. Since 1930 he worked at the Karpov Physicochemical Research Institute.
4   A.I. Akhiezer headed the Theoretical Department at the Kharkov Physical Engineering Institute in 1937–1964, Corresponding Member of the Ukrainian Academy since 1964. I.E Tamm was the Head of the Theoretical Department at the Physical Institute, USSR Academy of Sciences, Nobel Prize winner.
5   Formerly the All-Union Research and Design Institute of Power Engineering Technology.
6   At that time the reactor was coded as No. 7, and the whole facility, as Plant No. 3. After launching another heavy-water reactor, the facility was coded as No. 37.

CHAPTER 10

Natural Uranium Resources for the Nuclear Program

The pioneers of the extensive investigation of radioactive mineral resources were academicians V.I. Vernadskiy and A.E. Fersman. In 1939 V.M. Molotov, the Chairman of Sovnarkom (Council of People's Commissars) approved Vernadskiy's idea of setting up a Uranium Commission. By that time the Academy of Sciences had an Atomic Nucleus Commission headed by S.I. Vavilov. Vernadskiy proposed that the Uranium Commission should be assigned the authorities and finances to map potential uranium-bearing areas, survey and explore uranium deposits, and design technologies for uranium ore processing and uranium isotope separation. He also suggested that the Uranium Commission and the Atomic Nucleus Commission should coordinate their activities in the field of the research and utilization of uranium fission energy.

Not all of the outstanding scientists, especially physicists, supported the idea. In a historical novel [122] the author describes the attitudes of some outstanding physicists toward the creation of uranium industry. In early 1940, the first session of the newly created commission was held. It was attended by V.I. Vernadskiy, V.G. Khlopin, A.F. Ioffe, S.I. Vavilov, P.L. Kapitsa, A.E. Fersman, P.A. Svetlov (Scientific Secretary of the Academy of Sciences), two secretaries of the Commission, mineralogist D.I. Shcherbakov (VIMS) and geochemist L.V. Komlev (RIAN), and other scientists. Reporting on his meeting with Molotov, Vernadskiy noted that at that time Canada was actively mining uranium minerals and produced 120 g of radium annually, and that Belgium was actively exporting uranium ore from Congo. He said that uranium would bring an industrial revolution, and that the USSR should not fall behind in the utilization of uranium fission energy.[1]

Vernadskiy suggested that the new Commission should be headed by a physicist because physicists were the major consumers of radium and would be the major consumers of uranium. The comments of the academicians were not encouraging.

A.F. Ioffe: "We could hardly expect any practical benefit from uranium fission in the nearest future. I agree that more research work should be done in this field, but it is too early to consider the creation of uranium production industry to be an urgent problem."

S.I. Vavilov: "The uranium boom may in fact be a trick of some industrial companies. A demand for radium has been growing each year, especially because of the war. Uranium is a by-product in radioactive ore production. Its overstock may demand finding ways to get rid of it."

P.L. Kapitsa: "The dispute of how soon we will produce uranium energy depends on our efforts to master it. A rapid and efficient output requires much money and human and material resources. Being an engineer, I usually approach complicated problems from the engineering point of view."

Vernadskiy stated his opinion on the negative attitude of physicists toward the creation of the atomic industry in his address to V.G. Khlopin: "Vitaliy Grigorievich, I do not see any enthusiasm among our colleagues – physicists. So you will have to head the Uranium Commission." The Uranium Commission set to work rather quickly. Young professors – physicist I.V. Kurchatov, physicochemist Yu.B. Khariton, and geochemist A.P. Vinogradov were added to the Commission list.

Soon it became evident that our physicists-coryphaei were wrong, and the mineralogist Vernadskiy was right.

In late 1942 the USSR leaders became aware of researches conducted in Germany, Great Britain, and USA into the utilization of uranium fission energy for military purposes. As a result, a decision was made to mine uranium ore in the USSR. In November 1942 the State Defense Committee ordered the People's Commissariat of Nonferrous Metallurgy (Narkomtsvetmet) to proceed with the production of uranium from domestic ore. The Sovnarkom Committee on Geology was charged with prospecting of uranium deposits. In 1943, a Department of Radioactive Elements was included in the Committee, Regional Geological Surveys were set up, and a special Uranium Sector 6 was organized in the Fedorovskiy All-Union Institute for Mineral Resources – VIMS. VIMS was founded in 1918 as a head institute in prospecting, exploration, and development of deposits of uranium and ferrous, nonferrous, and rare metals. It provided resources for the defense industry in the Ural, Siberia and other regions. Future academician D.I. Shcherbakov, the well known follower of Vernadskiy, was appointed a scientific supervisor of a uranium sector (many deposits of rare metals were discovered owing to his predictions). The sector numbered more than 60 employees including a large group of scientists who were recalled from the Army. In a decree issued by the State Defense

Committee it was emphasized that uranium mining should be started in the deposits of Central Asia. In full conformity with the proposal previously made by Vernadskiy and Khlopin, the first action implemented by the Soviet government was accumulation of uranium ore reserves. In accordance with this decision, in 1943 mining and processing of uranium ore began at the Tabashar Mine, Tadzhikistan.

As early as 1930–1931, the GIREDMET Institute conducted research into extracting radium from the Tabashar ore. By 1935 a small settlement, a mine, and a hydrometallurgical factory were built. In 1941, Plant V of the Chief Rare Metals Directorate, Narkomtsvet Committee, and the Odessa Branch of the GIREDMET Institute were evacuated to Tabashar. The production of bismuth[2] and strontium salts and luminophor was set up here. In 1934, uranium ore was processed there to produce radium under the guidance of Professor I.Ya. Bashilov [50]. The Tabashar mine was laid up before the war. In her memoirs Z.V. Ershova recalls that as early as 1933, academician A.E. Fersman emphasized the good prospects of the Tabashar and Mailisu deposits discovered in 1926 and 1934, respectively.

AVRAAMIY PAVLOVICH ZAVENYAGIN,
Member of Special Committee.
Deputy Minister of Internal Affairs from 1945. Coordinated the construction of facilities of the atomic industry. Minister of the Atomic Industry from 1955 to 1957.

On December 8, 1944, the State Defense Committee decided to create a large uranium mining enterprise on the basis of the Tadzhikistan, Kirgizia and Uzbekistan deposits and, also, to transfer the management of this project from Narkomtsvetmet to NKVD. A decree issued by the State Defense Committee on May 15, 1945, set up NKVD Center No. 6 for mining uranium ore. Prior to the formation of the Special Committee and the PGU, 9th Directorate headed by A.P. Zavenyagin, Deputy of People's Commissar of Internal Affairs, was set up in the NKVD system. Zavenyagin was put in charge of uranium mining and the construction of various facilities intended for supporting the implementation of Program No. 1 and the researches conducted in the USSR Academy under the scientific guidance of Kurchatov. In a biographical essay devoted to Zavenyagin it is mentioned that he was engaged for solving the uranium problem in early 1943 [33]. The following dialogue took place in the Stalin's office: "Comrade Zavenyagin... You are a metallurgist and a miner. Do you know anything about the reserves of uranium and graphite?..." Zavenyagin answered that he had no information about the uranium ore, yet, he knew that there was graphite in Siberia, in the Lower Tunguska region, in the area of the Kureika River. A problem posed by Stalin was ultimately brief: "You should find both graphite and uranium and immediately start up mining." Zavenyagin got acquainted with Kurchatov in the same office. The Glavpromstroy Department of NKVD was also subordinate to Zavenyagin. This department was in charge of constructing the facilities of a future Center No. 6 in Central Asia, similar production centers in the Ural and a number of enterprises in other regions.

On October 1, 1945, Center No. 6 and the NII-9 institute were transferred from the NKVD to the PGU system. The first uranium resource base, Center No. 6, consisted of 7 mines and 5 plants, among which were:
- Plant V with a hydrometallurgical shop (Plant No. 4);
- Tabashar Mine (subsequently transformed into a mining department) – Enterprise No. 11;
- Adrasman Mine – Enterprise No. 12;
- Mailisu Mine – Enterprise No. 13;
- Uigur Mine – Enterprise No. 14;
- Tyuya-Muyun Mine – Enterprise No. 15.

Settlement Chkalovsk was built within 10 km of Leninabad (previously named Khodzhent). Later, the main hydrometallurgical plant of the Center was built there for processing uranium ore transported from the above-mentioned mines. The experimental hydrometallurgical shop in Tabashar (40 km of Chkalovsk) was reconstructed to Plant No.3[3] (director Ya.B. Slonimskiy). Here, apart from the Tabashar ore they processed the ore delivered from the Mailisu, Uigur, Adrasman and other mines. Ore transportation was among the most complicated problems.

P.Ya. Antropov recalled that in 1944–1945 uranium ore was brought to the processing plant through the Pamir mountain pathways in bags attached to donkeys and camels. Neither convenient roads nor appropriate engineering units and technology for processing uranium ore were available at that time. Conditions for mining uranium ore were extremely severe.

B.N. Chirkov, an NKVD colonel, who worked in 1944-1945 at the construction of the Dzhezkazgan copper-smelting works in Kazakhstan, was appointed the first director of Center No. 6 on March 7, 1945, on Zavenyagin's suggestion. At the end of 1942 Chirkov was Director of the Tyrnyauz tungsten-molybdenum plant in Kabarda-Balkaria. In 1942, he managed to evacuate the personnel and strategic resources via the Caucasus Range.[4] The extreme importance of mining natural uranium for creating an atomic bomb was emphasized by Stalin. Chirkov recalls in his memoirs that at the meeting on the occasion of his appointment Stalin said: "...Americans think that we will have an atomic bomb in 10–15 years, and their strategy is based on this assumption. Now they have only a few atomic bombs, yet, as soon as they equip their military air forces with atomic weapons, they will try to dictate us. This will take at least five years. By that time we must have our own atomic bomb. Comrade Kurchatov assured the Political Bureau that these terms are realistic. For the scientists, engineers and personally for you, comrade Chirkov, this problem is equivalent to the efforts of the war time from the standpoint of pressure and responsibility. You will be given whatever assistance you need and furnished with full powers. Your enterprise will have everything you need."

A.B. Dranovskiy was appointed Chief Engineer of Center No. 6. Later, this position was successively occupied by F.S. Vlasov, A.A. Popov, and P.I. Shapiro. M.F. Zenin was Deputy Director in geology since June 23, 1946, and since June 27, 1950, A.A. Daniliyants took this position. Because the role of builders was of great importance at that time, Director Chirkov was a supervisor of construction operations up to 1950. Since July 1, 1950, the position of Deputy Director in construction was successively occupied by M.M. Khaustov, K.V. Danilin, and A.A. Smolenskiy. The builders were mostly convicts, settlers, and soldiers. Their number was as great as 12,000.

Because of the lack of natural uranium supply for the PGU facilities under construction (experimental nuclear reactor F-1 in Moscow, production reactor in the South Ural, and gas-diffusion plant in the Middle Ural), the construction of Center No. 6 was a top priority. Maximum investments were assigned to the uranium mining enterprises.

It is pertinent to cite here Teller's note concerning the role of production of fissile materials [124]: "Production of fissile materials is the most difficult stage in creating the atomic bomb. As soon as a state finds a solution and implements the process, one can be sure that within a few months it will posses an atomic bomb."

The construction of facilities for production of fissile materials started in late 1945. In the subsequent years expenses for the construction of the atomic industry facilities increased ten to hundred times. The solution of problems specified in Program No. 1 called for the creation of new facilities in the PGU system as well as for the reconstruction of numerous plants and institutes from other branches of industry that worked for the atomic project. The amounts of state resources consumed by the atomic project was not known at that time to a wide community or even to all members of the government. A.P. Aleksandrov, President of the USSR Academy of Sciences, wrote over 30 years later [125]: "Only now can we openly and barely state that a major portion of difficulties experienced by our people during the first postwar years was demanded by a necessity to mobilize tremendous human and material resources to do our best for the successful and shortest completion of the scientific research and engineering projects aimed at the creation of nuclear weapons."[5] All those efforts were made to do away with the USA nuclear weapons monopoly and not to allow them to control the world community. Taking into account the extreme importance of the newly created industry, the government allowed construction operations to be started prior to the completion of design projects.

In the PGU structure the problems of uranium prospecting and mining and metallurgical and chemical production were of top priority.

Development of Production Center No. 6 and USSR Uranium-Bearing Areas

Enterprises of Center No. 6 placed under the control of the PGU started to be developed rapidly. In the fourth quarter of 1945, Plant No. 4 of the People's Commissariat of Nonferrous Metallurgy was reorganized and given an independent budget. To provide the building of uranium mines Nos 11–13, three contingents of builders were set up in the NKVD system mostly consisting of convicts and special settlers.

By the beginning of 1945, the Tabashar Mining Department was the only mining enterprise which was still in the exploration and preparation stage. By the end of 1946 the quantity of domestic uranium was insufficient even for a 50-percent load of experimental reactor F-1. In 1945, Center No. 6 managed to mine 18 thousand tons of uranium ore and produced 7 tons of uranium. In 1946 a 40-percent concentrate of uranium salts was produced amounting to 20 tons.

Until April 1946, all of the reconstruction operations were managed and controlled by Section 5 of the Engineering Technical Council. At its first session, which took place on January, 1946, the engineering project for the technological lines of Plants Nos 1–4, general plans for development of particular mines and Plant No. 4, and the design projects for mines Nos 11–13 were discussed. Among the members of the section were A.P. Zavenyagin, P.Ya. Antropov, S.E. Egorov, N.F. Kvaskov, V.B. Shevchenko, Professor V.I. Spitsin, V.S. Emeliyanov, E.D. Maltsev, top officials of the State Design Institute for Rare Metals of the Narkomtsvetmet (GIPROREDMET Institute, Director P.Z. Belskiy, Chief Engineer B.Ya. Bezymyanskiy). Besides the uranium separation technology to be applied at Center No. 6, Section 5 regularly considered many other problems. An example is a target assigned to the NII-9 Institute in the Section minutes of February 8, 1946:

– V.B. Shevchenko is to continue investigating of methods for extracting material A-9 (code name for natural uranium) by sulfuric acid, carbonate, and water leaching; recovery of A-9 from shale should be the main objective;

– study the possibility of combined recovery of A-9, molybdenum, nickel, and vanadium;

– draw up a project for constructing a mining-chemical enterprise for production of A-9 from Baltic shale by July 1, 1946.

The same resolution charged the NII-9 with drawing up a project, by May 1, 1946, for constructing a prospecting-mining adit and a pilot plant in Saka-Sillyame (now Sillamyae City, Estonia) with a production capacity for crushing, grinding, and roasting 100 tons of shale per day. The Section also decided to transfer Facility No. 3 belonging to the Voikov Plant (Moscow) to the PGU in 1946. This facility was designed for the semi-production testing of a technology for extracting uranium from shale. Yet, major problems arose in the development of Production Center No. 6. The PGU administration decided to rebuild the Tuya-Muyun Mine in 1946. They also decided to deepen Pit No. 2 at the Tabashar Mine in order to study the behavior of uranium deposits in primary mineralization zones. In 1945 the PGU 2nd Directorate was set up to provide specialists, equipment, and stocks for all of those operations, and to control the construction and rebuilding terms. This directorate included a mining and a metallurgical department (headed by Colonel D.D. Protopopov and Major-Engineer M.N. Sapozhnikov, respectively). A.A. Zadikyan, the former director of Technical Council of People's Commissariat of Nonferrous Metals was a chief engineer; scientists D.Ya. Srazhskiy and E.D. Maltsev and other leaders of the uranium industry worked for this directorate in 1947–1953.

The NTS Section 4 regularly discussed technologies applied in processing uranium ore. For instance, on October 11, 1946, they discussed a technology used in the processing of the Tabashar ore and a material balance of uranium for Production Center No. 6. It was emphasized that uranium loss in waste solutions was as high as 108 mg/liter because of an imperfect technology. In 1946 the average uranium recovery factor was as small as 56%. The NII-9 and institutes of the USSR Academy of Sciences were charged with improving the technological process.

After the defeat of Germany the Allies sent experts to the German territory to unravel the secrets of German scientists in production of nuclear weapons components and to look for the reserves of uranium in the occupied zones. In late 1945 uranium was brought from Germany to the Elektrostal Plant No. 12 located in the Moscow region. Here uranium bricks and even uranium slugs were manufactured for the experimental nuclear reactor F-1. The amount of uranium taken from Germany was insufficient for loading the production reactor, so Center No. 6 was assigned as a major supplier of uranium for this reactor. Among other consumers of natural uranium was a new plant which was to produce enriched 90-percent ^{235}U. It was rapidly constructed in the Verkh-Neivenskiy settlement, Central Ural. Another plant designed to produce enriched uranium by the electromagnetic method was under construction in Sverdlovsk-45, Central Ural. New nuclear reactors designed to produce plutonium were constructed in Chelyabinsk-40; the production of enriched uranium in Sverdlovsk-44 was enlarged. Besides, they planned to build new facilities, potential consumers of natural uranium, in other regions (Tomsk, Krasnoyarsk, and Angarsk). To provide domestic uranium for the production reactor, in 1947 the leaders of the PGU and Glavpromstroy Department of NKVD managed to put into service additional production facilities at Center No. 6, and the output of uranium ore was nearly 200 tons per day. Yet, this was obviously insufficient even for one reactor producing plutonium; the reactor had to be reloaded several times a year.

A decree issued by the USSR Council of People's Commissars on October 13, 1945, initiated uranium exploration in various regions of the country in order to expand uranium mining. A special Geological Exploration Department was set up in the Ministry of Geology. A period of the first 15 years was time of the formation of uranium geology. During this time several medium-size deposits were found in the Ukraine, North Caucasus, and Central Asia [126].

In addition to the Central Asia deposits, uranium-bearing iron-ore deposits were discovered in the Ukraine, Pervomaisk and Zheltorechensk, on the basis of which Production Center No. 9 was formed in 1951. Based on the Beshtau and Mt. Byk deposits discovered in the North Caucasus, the Lermontovsk Mine started to operate. In 1946 Production Center No. 9 mined 13.5 thousand tons of ore and recovered 13.4 tons of uranium.

Concurrent with the exploration of uranium deposits in the USSR, measures were taken to investigate the possibility of delivering uranium from East Germany and Czechoslovakia, where uranium was mined in Saxony and Jachymov in the 19th century. On November 23, 1945, a Czechoslovakian uranium industry was founded according to an agreement between the CSSR and the USSR. The intergovernmental agreement stipulated the further development of the Jachymov mines and joint exploration and mining operations. In Czechoslovakia they understood the USA monopoly of nuclear weapons as a threat to the peaceful co-existence of the world states. The nuclear research conducted by Soviet scientists prevented the violation of peace [127]. In October 1946, a similar governmental decree concerning cooperation with the DDR was issued. On May 10, 1947, a branch of the Soviet state joint-stock company "Bismuth" was set up in the DDR. A.P. Zavenyagin and P.Ya. Antropov contributed much to the creation of this company.

However, the amount of the ore delivered from the DDR and Czechoslovakia[6] and the extended production capacity of Center No. 6 still could not satisfy the increasing demand for uranium. Under the guidance of the Special Committee and PGU, attempts were made to speed up the production of uranium. It was suggested to use uranium-bearing shale, as well as the ore deposits intended for the ferrous metallurgy and coal industries, where iron and coal were mined. It was found that the iron ore contained uranium oxides. Examples were the uranium-rich iron ores of the Zheltorechensk and Pervomaisk deposits in the Krivoy Rog Basin. The Ministry of Ferrous Metallurgy was assigned to drive mines there for a combined production of uranium and iron. In 1951, the Zheltorechensk and Pervomaisk mines were placed under the control of the Second Chief Directorate (VGU) of the USSR Council of Ministers.[7] In 1947, it was found that some of the coal deposits located near the Issyk Kul Lake, Kirgizia, contained uranium. The Dzhil uranium-bearing coal deposit was surveyed and assessed. They proceeded with setting up an enterprise, Center No. 8, on the basis of this deposit. The technology of processing the Dzhil coal was to burn it up in the furnaces of the local thermal power station, and to process the ash at the hydrometallurgical plant. This coal-uranium enterprise was closed in 1956 as unpromising.

The amount of uranium mined in 1946-1949 (Table 14) was sufficient for the operation of the first production reactor, it allowed to proceed with construction of other plutonium-producing reactors, and provided enough uranium for producing highly enriched ^{235}U at a gas diffusion plant. However, the explored reserves of uranium were very small and poorly assessed. This fact was emphasized by P.Ya. Antropov as early as 1951 when the Production Center No. 6 was removed from the NKVD and placed under the control of the PGU: "... only the upper layers of deposits were studied, and the amount of deep mining and drilling was insufficient. As a result, the revealed reserves do not allow one to produce true estimates..." This situation remained for a long enough time.

The GIPROREDMET Institute was a major design contractor for the uranium industry. The leaders of the institute were the members of Section 5 of the Engineering Technical Council and, later, of Section 4 of the Scientific Technical Council. In fact the GIPROREDMET Institute was directly subordinated to A.P. Zavenyagin who emphasized that the structure of the atomic industry was different from the conventional mechanisms that were used in other industrial branches by the absence of the time factor. At that time the PGU administration reasoned that there was no need in creating its own institutes and plants: any enterprises of the country could be engaged for the atomic industry.

Table 14 Dynamics of explored reserves and production of uranium in 1946–1950.[*]

Year	Reserves, thousand tons			Uranium output, tons		
	USSR	East Europe	Total	USSR	East Europe	Total
1946	0.37	–	0.37	50	60	110
1947	1.43	0.34	1.77	129	210	339
1948	2.54	1.14	3.68	182	452	634
1949	3.97	1.87	5.84	278	989	1267
1950	5.50	3.22	8.72	417	1640	2057

[*] Exploration of uranium reserves was more active in the USSR, whereas uranium mining was more intense in Eastern Europe.

Besides the GIPROREDMET institute, the NII-9 was also involved in designing the uranium mines in the early period. A planning and design bureau (PKB) was set up at this institute. Its areas of activity were the Baltic region, the North Caucasus, and Production Center No. 8 in Kirgizia.

The Giproredmet institute was in charge of the development of Production Center No. 6. Project chief engineers were B.Ya. Bezymyanskiy, K.S. Kushenskiy, and D.S. Kutepov. A special design bureau (SPB-2) was set there with about 250 designers and draftsmen. It was headed first by M.F. Fedorovich, and later by N.S. Zagrebelnyi and T.F. Babkin. Since 1947, GIPROREDMET was overloaded with the PGU design projects, so a new design institute GSPI-12 was set up in Moscow especially for the atomic industry, in addition to the GSPI-11 institute.

By the end of 1948, Production Center No. 6 became the largest enterprise of the PGU. Its personnel numbered more than 15,000. For comparison these were the numbers of workers at other PGU enterprises:

- Plant No.12 (A.N. Kallistov, Director)   10,270
- Center No. 817 (B.G. Muzrukov, Director)   5,070
- Center No. 813 (A.I. Churin, Director)   3,546
- Plant No. 48 (P.A. Rastegaev, Director)   2,663
- Plant No. 544 (Sh.L. Teplitskiy, Director)   2,440
- Plant No. 906 (M.P. Anoshkin, Director)   673

Overall the PGU personnel in late 1948 numbered 55 thousand of full-time employees (without builders)[8] plus the staff of more than 100 collaborating organizations.

Because of a prime importance attached to the mining and processing of uranium ore a governmental decree of December 27, 1949, set up the Second Chief Directorate (VGU) subordinate to the USSR Council of Ministers. Production Centers No. 6 and No. 7[9], Mining Department No. 8 (Kirgizia), Plant No. 48 (Moscow), Plant No. 906 (Dneprodzerzhinsk), Severnoe and Ermakovsk Mining Departments, and a construction organization were transferred from the PGU to the VGU. P.Ya. Antropov was appointed Director of the VGU and N.B. Karpov, an experienced mining engineer, Hero of the Socialist Labor, the first Deputy Director. Doctor of Technical Sciences B.I. Nifontov was appointed another Deputy Director. Later in 1953–1965 he was Director of the GSPI-14 Institute. The VGU was charged with solving all problems that concerned the recovery of uranium from the ore to U_3O_8 production. The creation of the VGU substantially speeded up the operation of new uranium mining enterprises and sharply increased the output of natural uranium.

The production of uranium was continuously growing. As noted by foreign authors, in 1970 the USSR produced about 17,500 tons of uranium including 1,800 tons for peaceful purposes. In 1975 the world production of uranium (USSR production not included) made up 25,000 tons. Ores containing less than 0.2% U_3O_8 were not mined. However, in the case of an increasing demand for uranium this value may be as low as 0.02%. In 1974 the available world reserves of uranium contained in ore with a uranium concentration of about 0.1% were assessed as 4.5-5.0 million tons [84, 128].

These estimates varied greatly with the development of nuclear power industry. The uranium delivered to the USSR from East Europe was used to manufacture fuel elements for loading reactors at nuclear power stations in DDR, Hungary, Czechoslovakia, and Bulgaria. The rate of the growing capacity of the USSR uranium mining industry was so high that by 1990 the USSR accumulated large reserves of natural uranium, enough to sell it at the world market in considerable amounts. According to estimates made by the NUKEM Company (Germany), by 1991 the USSR had a stocks of about 200,000 tons of uranium, whereas all of the countries of market economy had 150,000 tons. In 1989, 7,730 tons of uranium were sold at the world market, and in 1990 − 15,800 tons. The lowest price of uranium, $19.2–21.7 for 1 kg of U_3O_8, was recorded in October 1990 [128].

In the 1960s the following enterprises were constructed:
- Vostochnyi mining and production center, Kirovgrad ore-bearing area, Ukraine;
- Lermontov mining department, Ergeninsk area, near Pyatigorsk;
- Caspian mining chemical center (MCC), Mangyshlak ore-bearing area, Kazakhstan;
- Navoi MCC, Kyzyl-Kum area, Uzbekistan;
- Tselinnyi MCC, North Kazakhstan ore-bearing area;
- Yuzhpolimetall, South-Kazakhstan ore-bearing area of Kirgizia;
- Malyshev mining department, Trans-Ural region;
- Argun MCC, Streltsovskiy ore-bearing area, Trans-Baikal region.

In 1945–1954 the most intensively developed enterprise was Production Center No. 6 located in Central Asia.

The amount of the explored profitable reserves of natural uranium in Russia with its tremendous potential reserves comes to only 25% of the total reserves of the CIS countries. These profitable reserves are concentrated in the Streltsovskiy ore-bearing area of the Trans-Baikal region (Fig.19) [126].

The structure of explored uranium reserves in Russia is summarized as of January 1, 1991:

Price of 1 kg of uranium:	below 80 roubles	26%
	80–120 roubles	20%
	>120 roubles	54%.

The 9-percent portion of the reserves explored at that time was recovered by underground leaching (Fig. 20), the 91-percent amount was mined in shafts. [Leaching of silicate and aluminosilicate ores was done using a solution of sulfuric acid and oxidants (pyrolusite, sodium or potassium chlorite, ferric iron ions)]. Also, widely used was open cut mining. This method was preferred in mining complex uranium ore at the Caspian and other mines (Tables 15 and 16).

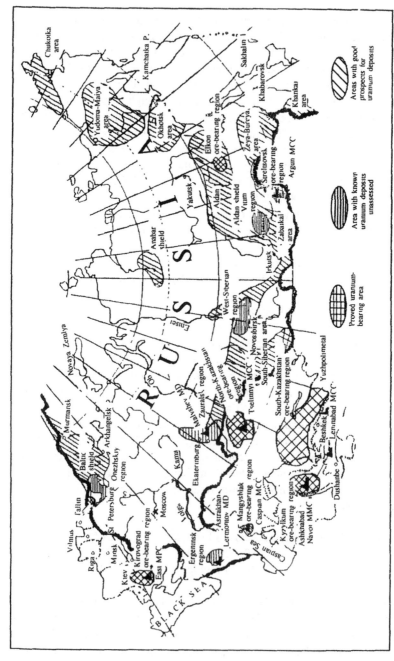

Fig. 19 Location of uranium-bearing areas and mining enterprises in the CIS countries, and promising uranium areas in Russia.

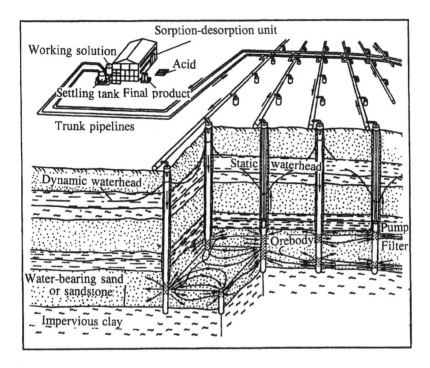

Fig. 20 **Technological schematic of an underground leaching site.**

In the USA, Canada, Australia and some African states uranium ore is mined in large quarries.

Classification and average yield of solid wastes at the mining and metallurgical facilities presented in Table 17 give an idea of growth of waste materials and the volume of waste piles around uranium mines, and also of the areas occupied by tailings piles near metallurgical plants. For comparison consider data on the amounts of waste produced in mining of other minerals [130]. On the average, 3 tons of waste per 1 ton of coal are produced in coal mining and about 0.3 ton of waste − during the consumption process. 5-6 tons of waste per 1 ton of steel are produced during iron ore mining, and 0.5-0.7 ton of waste − during the processing. In mining nonferrous metals, because of their low concentration in ore, no less than 100-150 and 50-60 tons of waste per 1 ton of metal are produced during the mining and processing, respectively. Up to 5-10 thousand tons and 10 to 100 thousand tons of waste are produced in mining and processing of 1 ton of rare, precious, and radioactive metals, respectively. From the early period of the atomic industry this drastic growth of waste products in the increasing utilization of rare and radioactive metals in the national economy posed a problem of devising low-waste technologies in the uranium-production industry.

Table 15 Comparative parameters of uranium ore mining by underground workings, open cut methods, and underground leaching [131].

Parameter	Mine	Quarry	Underground leaching
Permanent mining claim rated to ore-bearing area	1.2	1.6–1.7	0.1
Discharge of water into drainage system rated to static water reserves, %	13.8	21.5–22.0	–
Water discharge into tailings pile rated to static water reserves, %	22.0	23.0	1.1
Pulp accumulated in tailings pile	1.0	1.2	–
Dusted area	1.0	2.3	–
Landscape disturbance	Partial	Total	–
Uranium production in relation to reserves	1.0	1.25	1.4
Uranium recovery	0.70	0.81	0.88
Radon removal from mining and processing waste in relation to that in underground mining	1.0	1.2	0.03

Undeveloped uranium deposits of low economic value are known in the following areas of Russia:
- Baltic shield, Onega district;
- West Siberia;
- Aldan shield, Vitim and Elkon districts;
- Zeya-Bureya area;
- Khapkoiskiy area (south of the Maritime Territory);
- Okhotsk area;
- Yudoma–Maiya area, north of the Magadan region;
- Chukchi area;
- Anabar Shield (upper stream of the Khatanga River, Krasnoyarsk region).

Underground leaching is used to recover poor uranium ore seated at various depth under complicated geological conditions, particularly where underground mining and open cut methods are unprofitable. In the early period of the uranium industry, Academician B.N. Laskorin and Professor B.V. Nevskiy, veterans of the industry who contributed much to the development of the leaching method, noted: "Underground leaching is of growing economic value for it allows one to use highly leacheable ores which cannot be mined by conventional methods" [132, 133]. This method was not widely applied in Russia's deposits, but has to be developed. A technology for underground leaching of various uranium ores and other types of mineral deposits is described in detail in a monograph by Professor A.I. Kalabin [134]. A special arrangement of holes and pipelines for the delivery of working solutions to an orebody substantially reduces the amount of construction operations compared to shaft and open cut mining (Fig.20).

Table 16 Classification of quarries and mines in terms of ore output, in th. tons per year [132].

Size	Quarries	Mines*
Very small	below 100	below 20
Small	100–500	20–50
Medium	500–1000	100–250
Large	1000–5000	250–1000
Very large	5000	1000

* During the underground mining of 1 ton of uranium ore no less than 1.5–3 tons of waste are recovered to the surface.

Table 17 Classification of solid waste.

Source of waste	Type of waste	Yield rated to the total volume of rocks, %	Size, mm
Capital mine workings	Barren rock	3–4	−350...−500
Preparatory excavation	Marginal ore, barren rock	10–12	−350...−500
Conditioning workings	Tailings	3–5	−350...−500
Sorting at radiometric mill	Tailings	20–30	+50...−250
Leaching	Tailings	33–47	−15
Hydrometallurgical trade ore reduction	Tailings	17–22	−0.5...+0.043
	Sandy and silt tailings	32–47	−0.43

Although natural uranium is no longer used for manufacturing nuclear weapons, the number of economic uranium deposits in Russia is insufficient. In the opinion of experts there is a need for a detailed exploration of Russia's territory to provide enough fuel for nuclear power stations [126].

Uranium Ore Concentration

In 1937 Z.V. Ershova, a veteran of the uranium industry, visited the Tabashar Mine, together with professors V.G. Shcherbakov and D.I. Khlopin and other specialists from the People's Commissariat of Nonferrous Metals and USSR

Academy of Sciences, to inspect the mining and concentration of uranium ore at the Tabashar mine. "Our first impression of a chemical unit used to concentrate radioactive ore was desperate; the working conditions there were no better than in the shed laboratory of Marie Curie in 1898" [50].[10] Ershova was probably the first Soviet woman who visited a uranium shaft. She writes that the managers of the plant, geologists, and professor D.I. Shcherbakov tried to talk her out of going into the shaft. Ershova reclaims: "I joked that going underground cannot be more terrifying than going alone to Paris, put on a miner's overalls and took a lamp as did all miners of the group... ".

In the mine, the commission headed by Ershova witnessed a depredatory picking of richest ore. "Ore was picked using pneumatic drills, water was running down from the walls and ceiling. The air was full of radioactive ore dust... The ore was piled up on the surface near the exit from the adit. Here the ore was manually sorted. After this first manual concentration... the ore was carried in sample boxes to a Kolgerster electric gamma-ray meter to determine uranium concentration from gamma-ray radiation" [50]. Manual radiometric ore grading was used at the uranium mines of the USSR already in the prewar period.

In each deposit uranium was represented by a large group of minerals featuring diverse properties.[11] Where uranium is largely disseminated in non-uranium minerals, mechanical concentration yields neither good uranium concentrates nor low-uranium waste materials discharged to a tailings pile.

During the early operations, concentration of uranium ore was done following the experience gained in the work with other minerals [132, 133]. Basically the process of uranium concentration is aimed at (1) removing most of barren rock from the mined material and (2) separating the remaining ore material into products whose mineral composition is most favorable for subsequent technological operations. Cheap concentration methods applied in solving the first of these problems provided a profitable concentration of uranium ore. Solution of the second problem reduces the cost of hydro-metallurgical operations and facilitates a combined utilization of uranium ore (Table 17).

Tailings piles are usually located at the same sites close to the mines where barren rock and poor ore are dumped. Mine (low active) water and tailings of a hydrometallurgical plant (if located nearby) are dumped to the same site. Radioactivity is a general feature of diverse uranium-bearing rocks. The mined material passes through radiometric control stations as it is transported underground or on the surface in cars or any other containers. These stations register the recovered uranium, the removal of basic barren rock to the dump, and the separation of the ore into a specified number of grades differing in the metal concentration necessary for subsequent concentration. Ore grades comprising large amounts of barren rock are subject to the second phase of radiometric concentration.

Figure 21 presents a flow sheet of a radiometric mill providing the preparation of crushed uranium ore prior to its delivery to a hydrometallurgical plant.

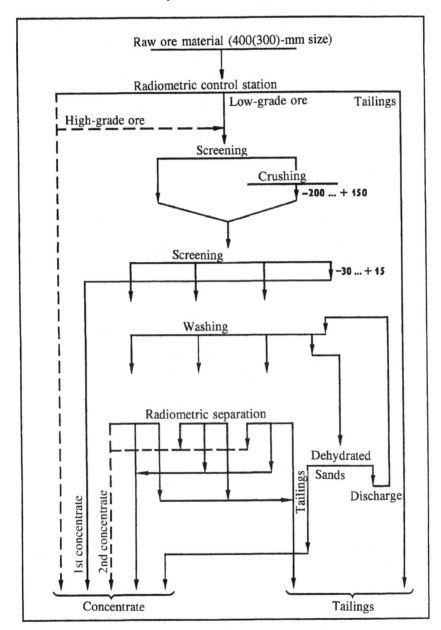

Fig. 21 Typical flow sheet of a radiometric mill.

The second phase of radiometric concentration includes the following preparatory operations: crushing (usually to a size of 150–200 mm), screening into 3–4 size classes, and the washing of classes to be radiometrically separated. The (−30...+15 mm) class usually is not subject to separation for in most cases it is characterized by an increased content of uranium [132].

The technological parameters of radiometric separation may vary over a broad range depending on the natural conditions of the ore, the sophistication of equipment, and the proper choice of technological techniques. Overall the yield of discarded tailings may be as high as 70–80% (sometimes more) of the total mass of the material brought for separation. All of the uranium production enterprises have mining geophysical services to monitor uranium content in various masses of ore under "in situ" conditions and after its picking and recovery to the surface. Using radiometric measurements of samples in mine workings and gamma-ray logs in wells and blast holes, these services promptly and cheaply outline orebodies and roughly appraise uranium reserves. Radiometric concentration is implemented with the aid of ore-grading units or separators. They consist of a radiation meter with a sensor and sorting (separating) devices, which deliver the ore material to the measurement zone and remove the separation products. This concentration procedure is constantly improved owing to advances in radioelectronic instrumentation and nuclear physics.

Who was in charge of solving these numerous scientific, engineering, and technological problems arising in uranium ore concentration and uranium oxides production?

After the establishment of the VGU under the USSR Council of Ministers, the scope of uranium ore mining and concentration was steadily growing. Concurrently with the creation of the GSPI-14 Institute, a new technological institute (NII-10, later VNIIKhT) was set up on the basis of 9 laboratories of GIREDMET on the initiative of the VGU managers and a governmental decree issued on April 17, 1951. A mining-geological and an electrochemical laboratory, 5 laboratories involved in the concentration of uranium ore, and a production analytical laboratory of physical analysis were transferred to the NII-10. The new institute was charged with solving the following problems:
- study the geology, mineralogy, and genesis of the most important uranium deposits of the USSR and the East European countries of People's Democracy;
- develop and manufacture geophysical equipment and deliver it to all operating uranium mines;
- devise a technology for concentrating low-grade ore;
- create highly economical flow sheets for hydrometallurgical processing of uranium ores and concentrates;
- study particular mining problems arising at uranium mines;
- devise the production analytical control and automation systems.

To solve these problems all necessary engineering services and research laboratories were set up in the institute. P.I. Buchikhin was appointed director of the institute with G.A. Meerson as a deputy director in science and A.A. Yakshin as a deputy director in geology.[12] According to a governmental decree of December 29, 1951, Facility No. 3 (experimental plant) was transferred from NII-9 to NII-10. N.S. Bogomolov was appointed director of this plant on April 17, 1951. According to a VGU order of February 6, 1952, the Podolsk experimental plant (with D.D. Sokolov[13] as director and G.E. Kaplan as deputy director in science) was assigned to take part in the NII-10 activities.

Crushing and grinding are the first preparatory operations in uranium ore processing at a hydrometallurgical plant. Usually to provide a better contact between the leaching agent and uranium minerals the material is ground to a size of 0.5 to 0.074 mm [132]. A required degree of grinding depends on the size of disseminated uranium mineral particles, ore porosity, and leaching agent activity. In the case of acid leaching, a lower grinding is allowed compared to carbonate (sodium) leaching. Uranium ore is crushed and ground in conventional crushers and ball mills. To control the crushing size conventional screens, graders, and hydrocyclones are used.

The procedure of hydrometallurgical uranium production was permanently improved [132, 133]. New methods for crushing and grinding were invented: crushing without balls and grinding in large-size mills (*Kaskad* mills); jet grinding; acid grinding without balls, i.e., combined with leaching. Ball mills were still in use too. As a rule, hydrometallurgical plants were equipped with a large number of various units. Because almost the whole of the ore delivered for processing was discharged to tailings piles (averagely it contained less than 1% of uranium), each year it was necessary to condemn new areas adjacent to the plant for arranging tailings piles.

The productive capacity of mills increases greatly when grinding metal balls are added. *Kaskad* mills of a large diameter and conventional ball mills operate in a closed cycle with spiral graders. Sinks of graders are subject to control hydrocycling and sands are brought back to ball mills.

As mentioned above, leaching with a solution of sulfuric acid and oxidants is a basic method for silicate and aluminosilicate ores. Ores with a high content of acid-consuming components (carbonates) are leached with solutions of sodium carbonate and bicarbonate. Minerals of tetravalent uranium are oxidized in a carbonate solution using a cheap oxidant – oxygen contained in the air.

A promising method for uranium recovery from "resistant" ores is autoclave high-temperature leaching which enhances the rate of useful reactions. The use of oxygen or air as an oxidant in autoclaves allows one to combine uranium leaching with the production of sulfuric acid owing to the oxidation of pyrite component in ores, as well as to oxidize sulphides of other valuable metals and dissolve them. Both oxygen and air are easily accessible oxidants; their application prevents the resulting products from contamination with any hazardous gaseous or dissolved materials. The application of autoclave leaching processes improves the extraction of uranium, saves chemicals, and reduces power expenditures (vapor, compressed air). The use of autoclave acid leaching facilitated the creation of a close-type ore processing flow chart.

Figure 22 shows a serial four-chamber autoclave equipped with mechanical mixers. This autoclave was devised in the USSR and is successfully applied in the atomic industry. There are horizontal and vertical autoclaves. Vertical autoclaves are equipped with pneumatic mixers, have a pulp volume of 100–200 m^3, and are intended to work at high temperature and pressure [129].

Fig. 22 Autoclave with mechanical pulp mixing. 1 – aeration and mixing device; 2 – mixer drive; 3 – autoclave housing.

A hydrometallurgical cycle for processing uranium ore includes a sorption recovery of uranium from the pulp and a subsequent refining of uranium solutions. In the early years when uranium ore was mined solely to produce uranium for defense purposes, associated products contained in the ore were completely ignored. Later, the situation changed and by-products were recovered on a commercial basis.

VNIIKhT was the leading institute of the industry charged with devising a technology for combined recovery of valuable products from ores. The researchers of name were A.P. Zefirov, B.V. Nevskiy, and B.N. Laskorin.[14] Under the guidance of these researchers the VNIIKhT developed and implemented a combined technology for producing uranous-uranic oxide (U_3O_8), thorium concentrate, and an ammonium phosphate fertilizer — ammophos, from uranium-phosphorous-thorium ores (Fig. 23). This figure demonstrates a crushing process during which minerals of uranium and thorium precipitate from a heavy suspension, a mixture of Fe_3O_4 and FeSi, while light-weight products, barren rock included, stay at the top. The resulting concentrate is stripped with sulfuric acid. Uranium is extracted from the phosphoric acid subsequent to its filtering through rotary-drum filters; the raffinate is ammoniated, evaporated, granulated, and dried up to produce ammonium phosphate fertilizer – ammophos. During the extraction, other products contained in the concentrate are dissolved after their decomposition and accumulate in organic and water components. Uranium and thorium are extracted from the phosphoric acid and separated in the course of re-extraction. The uranium re-extract is subject to additional extraction purification for producing crystals of ammonium uranyl tricarbonate which is roasted down to uranous-uranic oxide of subatomic purity. Thorium concentrate is precipitated from the thorium re-extract and stored until subsequent reprocessing.

It was for the first time in the world practice that uranium and all valuable associated components were extracted from low-grade, noncommercial ores of a complex composition with a simultaneous production of nitrogen-phosphoric fertilizers. The utilization of associated products gave a nearly twofold reduction of the cost of uranium [132]. The industrial implementation of this process eliminated environmental contamination with liquid wastes because process solutions are evaporated until a dry fertilizer is obtained. Gypsum produced in the reaction between sulfurous acid and ore is discharged into the tailings pile where it binds most of radioactive elements (radium). A group of scientists from the VNIIKhT Institute headed by Academician B.N. Laskorin, and researchers from other institutes devised and implemented combined ore processing technologies aimed at separating other metals. A technology designed for the recovery of rare earth elements and molybdenum along with uranium is described in detail in [133], where data on the extraction of molybdenum from thiosulfates by applying various sorbents are presented. It is shown that the extraction of uranium without a combined processing of low-grade uranium ores is absolutely unprofitable. At the same time the development of sorption-extraction processes resulted in the recovery of rare-earth and other useful elements (scandium, copper, etc.) and even in the building of mills near uranium mines producing gold and some other metals.

Environment Contamination and Protection

The mining and concentration of radioactive ores are unavoidably accompanied by environmental disturbances and harmful effects of uranium and thorium decay products on the vegetable and animal kingdoms. Isotopes ^{235}U, ^{238}U, and ^{232}Th are the fathers of three families of radioactive elements.

The father of the main family, ^{238}U decays to ^{226}Ra (half-life $T_{1/2} \sim 1600$ years) through five successive decay reactions. The latter continuously decays to ^{222}Rn ($T_{1/2} = 3.8$ days). A short-lived ^{222}Rn successively decays to other

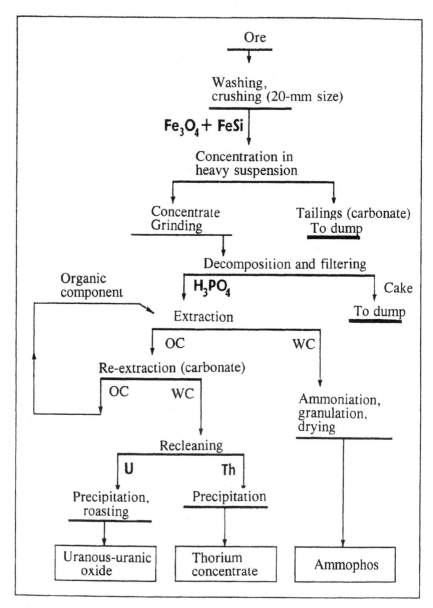

Fig. 23 Flow sheet for processing uranium-phosphorous-thorium ores. OC and
WC − organic and water components.

radioactive elements: first to a long-lived ^{210}Pb $(T_{1/2}=21$ years), and then to
^{210}Bi and stable ^{206}Pb. All of the successive decays (nearly 20) from ^{238}U to the

end product ^{206}Pb are accompanied by the emission of alpha-, beta-, and gamma-rays.

^{235}U, the father of another family, decays to a long-lived ^{227}Ac ($T_{1/2} = 21.6$ years) through a few successive decay reactions. The products of this series ^{223}Ra and ^{219}Rn have half-lives of 11.7 days and 4 s, respectively. ^{227}Ac decays to a stable isotope ^{207}Pb through 7 successive decays. ^{228}Th ($T_{1/2} = 5.8$ years) is a long-lived daughter product of ^{232}Th. The decay product ^{224}Ra has a half-life $T_{1/2} = 3.64$ days, and successively decays to a stable isotope of ^{208}Pb. Thus, due to the presence of above-mentioned radioactive decay products the radiation background in the underground uranium mines, quarries, and adjacent areas is a few times higher as compared to the natural one. In the case of poor ventilation, the content of gaseous radioactive products (basically of radon isotopes) and aerosols in the air of shafts and other premises of mines may be inadmissibly high.

Concentration of radon around uranium mines is more than ten times higher than its background value [135]. However, the major sources of atmospheric pollution and contamination of the areas adjacent to hydrometallurgical plants and uranium mines are aerosol ejections. As it follows from data presented by the Biophysics Institute (Public Health Ministry), the concentration of radioactive fallout produced by mining dust, vapor condensate, and other aerosols may be as great as 10^{-9} Ci/m$^2 \cdot$ day if measured in the total alpha activity units [135]. Usually these sites are small, and ejections are not very high (10–12 m). Along with dust and aerosol, basic products of uranium mining, uranium and radium, are removed to the environment, though ejections are usually dominated by alpha-activity of uranium (Table 18):

Table 18 Ratio of the alpha-activities of uranium and radium in settling dust at different distances from hydrometallurgical plants (averaged over all zones) [135].

Plant*	At produc-tion site	Distance from production site, km				
		0.3	0.3–0.6	0.6–0.9	0.9–1.2	1.2–1.5
GMZ-1	3.4	2.6	1.7	1.4	1.3	1.2
GMZ-2	18.0	10.0	6.5	3.7	3.4	2.8

* GMZ-1 produced only U$_3$O$_8$, GMZ-2 - uranium metal.

The established intakes of radium and uranium by vegetation are as follows: the content of radium in vegetation is no more than 1.2% of its content in soil; the uranium content may be as high as 20–25% (in radioactivity units)(Fig. 24).

Besides the radioactive elements belonging to the ^{238}U, ^{235}U, and ^{232}Th series, there are other elements of weak natural radioactivity (^{40}K, ^{129}I, ^{244}Pu, etc.). However, elements of the ^{238}U radioactive series, particularly ^{226}Ra and

²²²Rn, are the basic sources of a natural radioactive background and radiation fluxes associated with radionuclides.

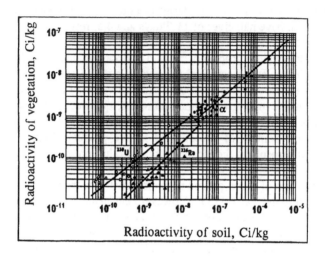

Fig. 24 **Relation between the radioactivities of soil and vegetation [135].**

Despite a high dispersion of uranium and, consequently, of its decay products in the Earth spheres (atmosphere, hydrosphere, and lithosphere), its concentration in rocks and soils varies greatly. This phenomenon is responsible for a variation in the radiation background over the Earth. For instance, a higher radiation is characteristic of areas where igneous acidic rocks (granites[15] and the like) are developed. Young sedimentary rocks feature higher radioactivity as compared to ancient ones because most of uranium in the latter has already decayed. Naturally, radiation background is higher in areas comprising uranium deposits as compared to adjacent and distant areas (Fig. 25). In the course of mining one can delineate a central zone, zones of primary and secondary effects, and an unaffected zone.

In some areas of the Earth radioactive elements accumulate close to the surface in high concentrations. Those areas feature highest natural radiation backgrounds. Vast areas exist where the radiation background is 10 and even 20 times higher than the average radiation background of the Earth. These are coastal placers of heavy sands with high concentrations of monazite which contains up to 10% of thorium and lesser amounts of uranium. Placers measuring tens and hundreds of square kilometers in area are known on the coasts of India, Brazil, Australia, and, in smaller areas, elsewhere.

The sources of pollution are areas of uranium quarries and adjacent areas where ore is piled up, radioactive barren rocks are disposed, and aerosols and dust are produced. Quarry waters containing compounds of uranium, radium, and other radioactive and heavy metals contaminate soil and water sources. Uranium and thorium are also contained in mine water and waste water of special laundries and shower-baths. ^{210}Po and ^{210}Pb — radium and radon decay products also present a particular radiation hazard.

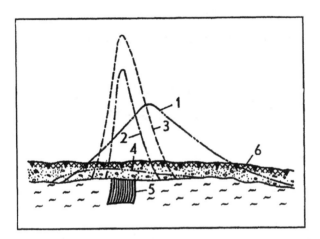

Fig. 25 **Relative intensities of radiation emitted by the major elements of the uranium series and the concentration of helium in soil above a uranium orebody [131]: 1–3 – radiation intensities of uranium, radium, and radon, respectively; 4 – concentration of helium; 5 – orebody; 6 – surface layer.**

Tailings piles consisting of solid and liquid components are formed in course of technological processes. The amount of solid waste is approximately equal to that of the ore, the amount of liquid waste being 2–3 times higher. The type of a tailings pile depends on the location of a mine and a hydrometallurgical plant.

In plain areas the tailings pile usually has an enclosing dam of waste sand around its entire perimeter, whereas in gorge-ravine areas the dam is built where the terrain lowers. Most common are alluvial tailings piles where the enclosing dam is built up by washing-in the tailings pulp pumped from the plant via a pulp line (Fig. 26). The enclosing dam of a bulk tailings pile is made of the solid, mostly sandy fraction, whereas the mud fraction is discharged into the resulting basin via a special pulp line together with waste water. By filling-in tailings pile is meant a dump where the pulp is pumped or poured into a drainless natural or artificial depression of the terrain (for instance, into an abandoned quarry).

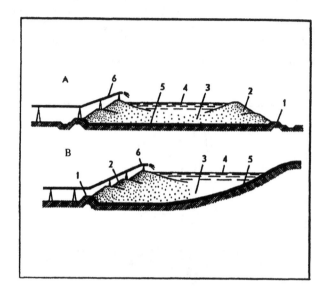

Fig. 26 **Design and basic components of alluvial tailings piles: A and B – plain and gorge-ravine tailings piles, respectively; 1 – forefront dam; 2 – alluvial dam (sand fraction of tailings); 3 – silt fraction of tailings; 4 – table of a settling pond; 5 – impervious seal; 6 – pulp line.**

The construction of a tailings pile crucially affects the extent and character of environmental radioactive contamination [135]. From the standpoint of contamination a tailings pile is differing from other industrial contamination sources. First of all this is a near-surface source of an irregular supply of radioactive aerosols having an extensive open surface (tens or hundreds of square kilometers) made up of fine-grained dusting materials (sand, mud). That is why the development of a tailings pile should include measures for reducing the area of dusting surfaces (slopes of enclosing dams, dried-up sites, etc.). For comparison Table 19 presents data describing the environmental radiation effects of various wastes of the uranium mining industry.

It should be emphasized that the efforts focused on providing sufficient natural uranium reserves for Program No. 1 ensured the creation of a nuclear shield for the Soviet Union and provided conditions necessary for the successful development of a nuclear power industry in Russia and in the adjacent countries of the former USSR. Suffice it to say that Kazakhstan, Russia, and Uzbekistan were among the ten largest world suppliers of uranium concentrate in 1992. These states produced 2,500, 2,000, and 2,070 tons of uranium, respectively,

Table 19   Characteristic parameters of the environmental insulation of recultivated objects.

Parameter	Mine and quarry dumps			Tailings piles of hydrometallurgical plants
	Barren (host) rocks	Radiometric concentration tailings	Marginal ore	
Concentration of uranium, %	0.001–0.006	0.010–0.020	0.010–0.030	0.001– 0.011
Total alpha-activity of rock, Bq/kg	1.4 · 10³– 4.4 · 10³	3 · 10³–10⁴	3 · 10³– 1.3 · 10⁴	1.03 · 10⁴– 4.9 · 10⁵
Specific density of surface the radon flux, Bq/m² · s	0.034–0.094	0.08–0.8	0.12–1.2	0.2–30
Concentration of ²²⁶Ra in rock, Bq/kg	37–220	300–550	300–1700	5000*
Gamma-activity of rocks, microR/h	20–120	40–250	50–200	100–1400
Approximate thickness of a recultivated layer needed for environmental insulation of wastes, m	1–1.5	1.5–2.0	1.2–1.6	2.5–3.0

* Average value (variation range of 1100 to 28000). A residue radiation does not exceed the admissible value of 0.2 Bq/m² · s at the specified thickness of the recultivated layer of disposed waste on the surface of radioactive objects covered with soil. This type of environmental protection is most widely used both in Russia and abroad.

which totals about 20% of the world production of uranium [138]. The present-day demand for natural uranium has reduced substantially. There is no need to produce plutonium and highly enriched uranium for nuclear weapons. The utilization of the already accumulated reserves of ^{235}U and plutonium, components manufactured of these radionuclides, and depleted uranium (dump) being stored at the uranium mills makes it possible not only to reduce crucially uranium mining, but also to stop it completely by the year of 2000 [139]. It is reported in [139] that during a period of 1938-1993, the world production of uranium amounted to 1.7-1.8 million tons. Its major portion is stored in the dumps of diffusion plants where the ^{235}U concentration is about 0.3%. Utilization of these dumps is equivalent to a surplus supply of 500-600 thousand tons of natural uranium for the nuclear power industry. In the case of utilizing weapons fissile materials there will be no need to increase the present-day production of uranium after 2010.

NOTES

1   In the USSR radium was basically used by physicians in treating oncological diseases and by physicists in neutron physics research. Neutron sources were

produced by the RIAN Institute from radium and its fission product, radon.

2 Subsequently, bismuth was used to produce polonium by irradiating it with neutrons. Up to now ^{210}Po and beryllium are used in production of neutron sources.

3 In 1948 Plant No. 3 was reconstructed again and was known as Plant No. 6.

4 After B.N. Chirkov was transferred in August 1953 to Zheltye Vody in Ukraine as a director of a production center, D.T. Desyatnikov, G.V. Zubarev, and V.Ya. Oplanchuk acted successively as directors of Center No.6.

5 In some years, up to 300-350 thousand builders and erectors were occupied in creating the atomic industry.

6 Later, the mining of uranium ore was started in Bulgaria and Poland.

7 Since 1949 all of the uranium mining industry was no longer under the PGU subordination.

8 The number of builders working at the sites of the atomic project in 1948 was higher than value by a factor of nearly 4.

9 Center No. 7 was a code name for the Saka-Sillyame enterprise, Estonia.

10 In 1932–1938 the laboratory was headed by its founder, Professor I.Ya. Bashilov who was also a scientific supervisor of an experimental Giredmet plant designed for uranium ore concentration.

11 Uranium is a constituent of more than 150 minerals along with alkaline-earth, rare-earth, and heavy metals [133].

12 Since January 1957 A.P Zefirov was director of NII-10 with P.I. Buchikhin and D.Ya. Surazhskiy as deputy directors in science and geology, respectively.

13 Later he worked as a chief engineer of the Elektrostal Plant No. 12, as a deputy of the chief engineer of Chief Directorate No. 16, and as a deputy director of the Minsredmash Scientific and Technical Department.

14 He joined VNIIKhT in 1952. Before that, he worked for institutes of chemical industry, a winner of the Lenin and State prizes.

15 As follows from data presented by RIAN, in St. Petersburg the radiation near the granite pillars of the Isaakiy Cathedral and near the Rostral pillars of the Vasilievskiy Island is 15 and 100 times higher than the natural radiation background, respectively.

CHAPTER 11

First Plant for Producing Items of Uranium Metal and Other Materials for Atomic Industry

It was planned to produce about 200 tons of uranium metal for the F-1 experimental reactor and the first production uranium-graphite reactor in 1946–1947. Nearly 150 tons of this uranium were intended for manufacturing cylindrical slugs, 35 mm in diameter and 100 mm in height, tightly encased in aluminum jackets. Plant No. 12 was the first enterprise that mastered an industrial technology for producing uranium metal and manufacturing items of this metal. The plant was a pioneer in the industrial production of uranium slugs, the items that were used from 1945 in nuclear reactors at Chelyabinsk-40 for producing plutonium for nuclear weapons. Without the efforts of Plant No. 12 it would have been impossible to produce on time another nuclear explosive, highly concentrated ^{235}U (at Center No. 813, Chelyabinsk-44).

Plant No. 12 was the cradle of nuclear and thermonuclear weapons: a technology for manufacturing dispersion slugs from enriched uranium and a technology for lithium production were devised there. Lithium and tritium derived by irradiating lithium with neutrons were the most important materials without which one could not create effective thermonuclear weapons.

By 1945, Plant No. 12 already had a 30-year history of producing defense armament. The plant started to be reconstructed immediately after it was placed under the PGU control. This was necessary because in early 1946 the plant was charged with a top secret job to begin processing uranium ore for producing specified amounts of uranium dioxide and uranous-uranic oxide [140]: "Our task was to produce uranium metal by conversion through the reduction smelting of uranous-uranic oxide together with sodium metal. Uranium metal was to be produced in the form of rods which should be machined, cut into small blocks, and sealed into aluminum jackets." This demanded the reconstruction of many workshops and production sites and the recruiting of new specialists.

The reconstruction of the plant was preceded by extensive research work aimed at devising technological processes to be installed at the plant as required by the needs of atomic industry.

Contribution of the GIREDMET and NII-9 Institutes

These institutes were pioneers in designing technological processes that were implemented at Plant No. 12.

The State Institute of Rare Metals (GIREDMET), People's Commissariat of Non-Ferrous Metals, was set up in Moscow in 1931. This institute started implementing projects of Program No. 1 in 1943, almost simultaneously with Laboratory No. 2. A.P. Zefirov[1] was Director of the GIREDMET institute and N.P. Sazhin (Academician since 1964) was Deputy Director in research. In 1943 the GIREDMET was charged with restarting the previously suspended activity of the Uranium-Radium Laboratory and including its research into a top-priority program.

The GIREDMET institute was charged with:
- updating technology for production of uranium from ore;
- production of uranium tetrafluoride;
- production of uranium carbide;
- production of uranium metal;
- designing and manufacturing diverse uranium targets to be irradiated with neutrons in various physical studies;
- production of diverse compounds of the uranium fissile products.

In early 1944 the institute produced uranium carbide for the first time in the USSR. Deputy Director N.P. Sazhin and Z.V. Ershova, a researcher, personally delivered this material in 10-kilogram portions to Laboratory No. 2, to I.V. Kurchatov and V.V. Goncharov for examination and preparation for its utilization in the experimental reactor F-1 which was under construction [24]. Apart from N.P. Sazhin, Research Director, the team of researchers who developed the first experimental technology for producing uranium metal included Z.V. Ershova, G.E. Kaplan, G.M. Komovskiy, Yu.A. Chernikhov, E.A. Kamenskaya, etc. M.G. Pervukhin, Deputy Chairman of the Council of People's Commissars, A.P. Zavenyagin, Deputy People's Commissar of Internal Affairs, and a Major-General S.E. Egorov, future director of the 2nd Department of the PGU, attended the experimental production of the first uranium metal bar (experiment was concluded at 2:00 a.m.). Thus, M.G. Pervukhin was the top man who organized the work under all projects of the nuclear program and formed appropriate teams in the People's Commissariats and at the institutes and plants engaged for the creation of nuclear weapons, prior to the organization of the Special Committee, headed by L.P. Beriya, and the First Chief Directorate (PGU). Together with Kurchatov, research director of the problem, and Zavenyagin, he was at the cradle of designing and testing the first atomic bombs manufactured from domestic plutonium and ^{235}U and creating the first thermonuclear bomb. Under his guidance the production of deuterium and lithium at chemical plants and tritium at PGU enterprises got under way.

Uranium slugs for nuclear reactors and diffusion filters for the equipment used to produce enriched ^{235}U were the basic components and materials for manufacturing nuclear weapons.

The successful production of uranium metal by Soviet scientists attested that Soviet specialists were capable of creating a uranium production industry and proceed with the implementation of the atomic program. To this end, an experimental facility was built at Plant No. 12. Here in 1945, semi-commercial experiments on the production of uranium bars weighing several kilograms were conducted. These experiments were supervised by a team of GIREDMET researchers headed by Z.V. Ershova.

One of the GIREDMET projects was the production of thorium. New laboratories were set up which needed additional premises. Under the guidance of Professor G.F. Komovskiy, X-ray diffraction and spectral analyses were organized at the institute. Professor Yu.F. Chernikhov organized a chemical analytical laboratory, and Professor G.F. Meerson — a uranium carbide laboratory.

ALEKSEY PETROVICH ZEFIROV, Corresponding Member of the USSR Academy of Sciences from 1968. From 1943 to 1946 directed the State Institute of Rare Metals where uranium carbide (1944) and first uranium metal bar were produced. Coordinator of Scientific and technological projects of the Atomic Industry from 1946.

In 1944 researches were under way on the production of uranium dioxide and the vacuum refining of uranium metal. The steadily increasing amount of uranium research conducted at the GIREDMET, and also Ershova's proposal in her memorandum to Zavenyagin on the necessity of creating a new institute in the NKVD system, promoted the formation of the NII-9 Institute (headed by V.B. Shevchenko) in November 1944.

In 1945 the NII-9 institute was charged with all uranium projects that were previously handled by the GIREDMET. The institute's tasks were formulated as follows:

VICTOR BORISOVICH SHEVCHENKO.
From 1945 to 1952 directed the NII-9 Institute dealing with the production of uranium and pluto-nium metal, and plutonium com-ponents of nuclear weapons.

- to investigate the known uranium deposits, systematize all data, and create a mineralogical museum;
- to develop techniques for uranium exploration and mining;
- to devise a technology for processing uranium ore from crushing to the production of pure uranium salts;
- to devise a technology for uranium metal production;
- to verify a technology for the extrac-tion of plutonium from uranium irradiated in nuclear reactors and technologies for producing plutonium metal and plutonium items.

In early 1945 NII-9 was charged with implementing a project for the separation of uranium isotopes by a centrifugal method. To this end, a special laboratory was set up headed by Professor F.F. Lange. Though this laboratory formally was included in the NII-9, its activity followed plans approved by the PGU.

In addition, the NII-9 institute was charged with some other problems, for instance, the production of heavy water.

Academician V.G. Khlopin comprehensively examined a technical project for the construction of the institute premises and a scientific program placed before the institute. He was surprised by a large number of problems posed, but still approved the program.

It should be noted that other academic and specialized research institutes and design bureaus were already charged at that time with solving some of these problems in accordance with the instructions given by the State Defense Committee, 9th Directorate of NKVD, and PGU. For instance, RIAN was a head institute in developing a plutonium recovery technology for a radiochemical plant; NII-42, People's Commissariat of Chemical Industry, was charged with production of uranium tetrafluoride; GEOKhI was responsible for the chemical and analytical control; Laboratory No. 2^2 — for construction of a commercial reactor and production of ^{235}U. At the department of inorganic chemistry, Moscow State University, various compounds of uranium were studied under the guidance of professors V.I. Spitsin and V.V. Fomin. Yet, most of the problems involved in the creation of technological processes to be implemented at Plant No. 12 were handled by the NII-9 laboratories.

In April 1944, V.B. Shevchenko and I.N. Golovin, representative of Laboratory No. 2, were sent to Austria (to the Radium Institute) for studying the possibility of annexing some equipment and chemicals. Besides, commissions consisting of A.P. Zavenyagin, V.B. Shevchenko, V.V. Goncharov, I.K. Kikoin, Yu.B. Khariton, and others were sent to Germany to decide on the possible evacuation of natural uranium (uranium salts and ore), as well as of various instruments, units, and equipment from Germany. One of the commissions arranged invitations to German scientists and skilled specialists to work at Soviet enterprises on a contract basis [89].

V.B. Shevchenko, Director of NII-9, aided by Z.V. Ershova and V.D. Nikolskiy, who were transferred from the GIREDMET, and officers of the PGU Personnel Department, staffed the research institute, formed the teams of analysts for its laboratories, and teams of researchers and engineers for various departments and experimental facilities.

The PGU managers and Zavenyagin personally gave much attention to the development of NII-9: to its projects, personnel, and experimental units. On January 17 and February 8, 1946, the Mining and Metallurgical Section of the Engineering and Technical Council considered the basic problems concerning the NII-9, Plant No. 12, and uranium mines. German specialists Dr. N. Riel and Dr. G. Wirths and Yu.N. Golovanov, Chief Engineer of Plant No. 12, took part in this session as experts. They submitted a conclusion on a project for the reconstruction of the plant workshops. Having examined the projects for the reconstruction of the concentration, purification, and uranium metal production workshops (Nos 2 and 3, respectively), the Section adopted an extended resolution covering the work of NII-9, and of Plant No. 12 in particular. The following assignments were produced:

- Provision should be made for the production of uranium dioxide at Workshop No. 2 without disturbing its equipment layout. The reduction of uranous-uranic oxide to uranium dioxide should be conducted in a separate building.
- A process for the recovery of radium from the crude sulfate and radium concentrates produced at Workshop No. 1 should be developed. The Ukhta Center of NKVD should be recruited for this project.
- Flow sheets for the workshops of Plant No. 12 and the product recovery percentage should be discussed within 10 days jointly with Academician V.G. Khlopin, German scientists Dr. N. Riel and Dr. G. Wirths, and also with V.I. Spitsin, V.B. Shevchenko, N.F. Kvaskov[3], N.P. Sazhin, O.E. Zvyagintsev, and some other scientists.
- V.B. Shevchenko, Director of NII-9, should arrange testing of all technologies of Plant No. 12 at the NII-9 laboratories within three months.
- GIPROREDMET (P.E. Belskiy) should arrange installation of special filters in the design of scavenging equipment to provide trapping of uranium from waste water.

In addition to these, the following assignments were made:

- V.B. Shevchenko should investigate the possibility of recovering uranium, molybdenum, nickel, and vanadium from shales.
- A plan should be drawn up by May 1, 1946, for mining the Saka-Sillyame shales and constructing an experimental plant using the project prepared by the VIMS Institute. The production of the plant's shale crushing, grinding, and roasting operations must be as much as 100 tons per day.
- Appeal to the USSR People's Commissariat to transfer experimental facility No. 3, belonging to the Voikov Plant, Moscow, People's Commissariat of Nonferrous Metals, to the PGU to conduct semi-commercial tests of technologies for uranium recovery from shale.

This illustrates the huge amount of work done by NII-9 in 1946. The institute had to answer for many projects, and also got much help. For instance, in early 1946, Facility No. 3 was transferred to the institute and became its experimental plant.[4] N.S. Bogomolov was appointed first director of Facility No. 3; subsequently, he was replaced by A.A. Matveev. By mid-1946 the personnel of NII-9 numbered 1200. The institute included 13 laboratories, a design bureau with its Leningrad Branch, and all necessary logistics services. A special design bureau was set up for the German professor M. Faemer dealing with the production of heavy water. Later, the NII-9 projects were somewhat revised.

To adjust the technology destined for the radiochemical plant that was under construction at Center No. 817, a special installation, No. 5, was constructed in NII-9. Here researchers from NII-9, RIAN, IONKh, IFKhAN, LIPAN, and other institutes adjusted a control system technology and equipment to be used for separating plutonium from uranium slugs irradiated in nuclear reactors. These operations were conducted under the scientific guidance of B.A. Nikitin, Corresponding member of the USSR Academy of Sciences, Deputy Director of RIAN. M.E. Pozharskaya, a veteran of NII-9, recalls [38] that the first director of this installation (belonging to Ershova's Laboratory) was P.I. Tochenyi, and after he left for Center No. 817 to become the director of the radiochemical plant, his post was successively held by M.A. Belokurova and M.V. Ugrumov. Under the guidance of Z.V. Ershova and V.D. Nikolskiy, on December 18, 1947, the NII-9 researchers managed to separate the first specimen of trivalent plutonium, 73 μg in mass. That was a light blue droplet placed in a millimeter test tube [38]. During the early half of 1948 two large batches of plutonium concentrate comprising 1,207 and 2,649 μg of plutonium were produced. Certain amounts of this plutonium concentrate were delivered to other institutes for research: 73 μg to Laboratory No. 2, 44 μg to IONKh, and 34 μg to RIAN. Most of the plutonium was used at the laboratories of NII-9 (research groups headed by V.D. Nikolskiy, V.S. Sokolov, and Ya.M. Sterlin). The group headed by Ya.M. Sterlin produced first plutonium metal; its properties were studied in the laboratory headed by A.S. Zaimovskiy. By the end of 1948, the

NII-9 institute had at its disposal nearly 300 mg of plutonium extracted from the concentrates produced on installation No. 5. This amount was sufficient to proceed with the final testing of the plutonium refining technology proposed and developed by Academician I.I. Chernyaev and his team [38].

Among the prominent scientists working at NII-9 was S.T. Konobeevskiy, Corresponding member of the Academy. In 1948 he was transferred from the Lomonosov Moscow State University to head a group of X-ray diffraction analysis. He was the first to study the phase state diagrams of uranium and plutonium alloys. In 1952 S.T. Konobeevskiy was appointed Deputy Director of NII-9 in research by a decree of the Council of Ministers. He is rightly considered to be a creator of the Soviet school of radiation materials technology. On December 1, 1952, Academician A.A. Bochvar was appointed Director of VNIINM institute with V.V. Fomin, professor of the Moscow State University as first Deputy Director.

Reconstruction of Plant No. 12

Initially Plant No. 12 was a relatively large enterprise with a personnel of more than 4,500. There were more than a hundred small buildings ranging between 150 and 500 m^2 in area, where various workshops and areas designed for charging conventional munitions were located. A change in the plant profile and development of new technological processes called for setting up a special building organization and, also a chief metallurgist and a chief technologist.

To arrange the reconstruction of the plant workshops and build new ones, in March 1946 A.N. Komarovskiy, Director of the Glavpromstroy Directorate, issued an order which appointed colonel-engineer N.N. Volgin, an experienced military builder, as Director of the NKVD Construction Directorate. In his memoirs [141] N.N. Volgin describes numerous episodes that occurred during the war and at construction sites of the Glavpromstroy Directorate, NKVD. Apart from Plant No. 12, Volgin mentions that he constructed other facilities of the atomic industry. When L.P. Beriya visited Plant No. 12, Volgin informed him of his activities.

In July 1949, by an order of the USSR Ministry of Internal Affairs, Volgin was appointed First Deputy Director of the Glavpromstroy Directorate of NKVD. In 1953 to 1974, after the Minsredmash Ministry was organized, he worked as a Director of the 10th Construction Directorate which in the early period was known as the First Chief Construction Directorate of Minsredmash Ministry. Construction sites under that Directorate were located in the Ural and Siberia. Besides, Volgin headed the construction of some objects relating to other industrial branches, e.g., the Ministry of Oil and Chemical Industry (Angarsk and Omsk), Siberian Branch of the Academy (Novosibirsk); construction of missile starting sites and a uranium mining center in Krasno-kamensk.

ALEKSANDR SEMENOVICH ZAIMOVSKIY,
Corresponding Member of the USSR Academy of Sciences from 1958. Head of the Metal Technologies Laboratory dealing with production of metal plutonium, uranium, zirconium, and alloys of these metals. Deputy Director of NII-9 from 1960 to 1977.

A project for the first reconstruction of Plant No. 12 was adopted by a People's Commissariat Council's decree of October 13, 1945. By 1956 six projects of the complex reconstruction of the plant were implemented.

The most important objective placed before Plant No. 12 in the early period was devising a commercial technology for fabricating metal uranium slugs to be loaded in the first production uranium-graphite reactor. A 35-mm uranium core should contain any materials of high neutron absorption (boron, rare-earth elements, etc.). This core was hermetically canned in a 1-mm aluminum jacket. When placed in the operating nuclear reactor it was cooled with chemically pure water flowing at a speed of 5 m/s through the slug-tube annulus. The uranium slugs were loaded into nearly 1,100 fuel channels, 75 slugs into each of the channels. To produce a specified amount of plutonium, the uranium slugs should be irradiated in reactor during 3 to 4 months without any changes in their sizes and failure in tightness. To this end, nearly 50 tons of uranium slugs (stripped) should have been produced for Laboratory No. 2.

First technical specifications to the fabrication of uranous-uranic oxide and metal uranium for F-1 reactor were devised by I.V. Kurchatov in 1944, and those for metal uranium slugs, on February 22, 1946 [24]. A decision to start commercial metal uranium production at Plant No. 12 was taken in late 1945.

First problems arose from the outset. " ...Boron contained in graphite molds contaminated uranium during casting. A boron contaminant also evolved from the enamel which covered technological vessels. Uranium was also contaminated by ferrous components contained in metallic calcium and used to reduce uranium" [24]. Most of products were defective. In 1945 Kurchatov informed the government (Stalin) of the situation with the production of pure uranium: "Though the problem of complying with the requirements for the purity of uranium metal is rather difficult, it undoubtedly can be solved. In my opinion, the convincing fact is that even in the case of the more crude production of graphitized electrodes we managed to achieve a contamination level which is lower than that for uranium metal at least by a factor of 20." In his memorandum Kurchatov did not mention anyone personally; he emphasized instead that the production of uranium metal was unsatisfactory because of a poor management of this project [24]. Updated specifications on a batch of metal uranium slugs to be produced at Plant No. 12 for F-1 reactor required control of admissible impurities of 14 elements. No limits were set up for other elements for lack of appropriate analytical control methods.

Kurchatov checked the presence of boron personally. The PGU chiefs B.L. Vannikov and P.Ya. Antropov supplied all necessary materials for the plant. For instance, calcium metal was delivered by aircraft from Germany (Biterfeld) because in 1946 the plant could not produce its own calcium [142]. The first emergency situation caused by the contamination of uranium by boron took place in November 1946 when it was found that the concentration of boron in one of the batches of uranium slugs destined for F-1 reactor was nearly 100 times higher than the specified value. Everything was checked up: the calcium delivered from Germany, the graphite pots for uranium smelting, and the graphite moulds for the metal casting. It was found that the Moscow graphite plant (Director S.E. Vyatkin) had shipped moulds of nonchlorinated graphite.

As mentioned above, an Analytical Council was set up in the PGU on October 26, 1946. To make its sessions more effective each analytical laboratory engaged by the PGU delegated a representative to the council membership. Among its full-time members were eminent scientists such as I.P. Alimarin, D.I. Ryabchikov, and I.V. Tananaev. A regular day was set up for the council sessions – each wednesday, at 5.00 p.m. The Analytical Council was mostly concerned with providing Plant No. 12 with uranium production techniques.[5] A Central Laboratory (TsZL) was set up at the plant. One can judge the hard work of the TsZL analysts from the number of analyses they made at that time (Table 20).

Table 20 Activity of the Central Laboratory of Plant No. 12 in 1947–1951.

Year	Personnel	Number of analyses, thousand		
		Chemical	Spectral	Total
1947	199	63.8	31.1	94.9
1948	231	51.5	29.8	81.3
1950	263	62.3	36.2	98.5
1951	263	73.1	31.5	104.6

Upon eliminating many problems, they succeeded in implementing the production of uranium metal slugs at Plant No. 12. In 1946, they manufactured 19 thousand slugs, 32 and 35 mm in diameter (overall weight of 36 tons), and 9 tons of balls, 80 mm in diameter, of uranium dioxide destined for F-1 reactor in full compliance with technical specifications. On December 25, 1946, a controlled nuclear chain reaction was initiated for the first time in Europe and Asia.

Where did they get uranium for Plant No. 12 in 1945–1948 and how did they manage to reconstruct the plant?

The first uranium slugs were made of crude uranium dust which was brought from Germany. In the first half of 1947 a workshop designed for processing imported high-grade uranium ore was put into operation at Plant No. 12. Domestic low-grade ores were reprocessed at Plant No. 906, Dneprodzerzhinsk (Director M.P. Anoshkin) since 1947 [142].

More than 20 buildings were built and reconstructed for the production of materials for the atomic industry (Table 21).

In 1946, the personnel of the plant numbered ca. 4,600, including ca. 700 engineers and technicians (100 technicians had higher education). In the early period there were many scientists and engineers from GIREDMET, NII-9, other institutes and, also, from Germany. The German specialists organized the production of uranium metal by the calcium-based thermal reduction of uranium oxide. The technology was imperfect, and researchers of NII-9 (laboratory headed by Professor A.N. Volskiy) and A.G. Samoilov had to improve it. In 1946 a technology for calcium-based thermal reduction of uranium from UF_4 was implemented at the plant.

Several emergency situations had occurred until they managed to adjust the technology for producing uranium metal. A.G. Samoilov, a veteran of the atomic industry, recalls [38] that in a workshop headed by N.N. Sinichenko uranium was produced by a method proposed by German researchers. "Uranous-uranic oxide (shavings) was reduced by adding calcium chloride to the stock (to provide slag formation). The entire mass of the stock was heated up in a hermetically sealed metal vessel and then chemically leached to produce uranium metal powder." Concurrently, Chief Metallurgist Yu.N. Golovanov conducted experimental melting to produce uranium metal by reducing uranium tetraflouride with the use of distilled calcium. In the course of one melting an accident occurred: " ...after a long, increasing roar, the bonnet of a heated-up vessel was torn off from the secure pins, flew away through a window and fell down a few tens of meters from the workshop building. The fire mass that flew out of the reactor caused a fire in the workshop. All of us took part in putting out the fire until firemen arrived. Since that time we were more careful. These meltings were conducted in the same workshop every three to five days within many months."

In the production of uranium metal of great importance was the manufacture of a durable melting crucible. There was a special ceramics laboratory at the NII-9 headed by S.G. Tresvyatskiy. Researchers from the NII-9 metallurgical laboratory were also involved in improving the crucible. A technology for lining the inner surface of a crucible consisting of a chiselled casing (Armco iron) with magnesium or calcium oxide was developed by F.G. Reshetnikov, A.G. Samoilov, and M.A. Dunskiy. A hazardous accident took place during high-temperature smelting of uranium: "In the course of an experimental smelting the bonnet of the vessel was torn off with a terrible crash, high-temperature smelting products were ejected, struck the ceiling, and welded on it. Luckily, nobody suffered."

Table 21   Development of production capacity at Plant No. 12.

Process	Number of main buildings	Years	Production area, thousand m^2	Investments, mln roubles
Manufacture of uranium articles		1946–1967		13.7
experimental	1		5.4	
commercial	4		15.1	
Uranium ore processing	4	1947–1957	25.9	11.0
Calcium production	4	1948–1967	18.0	4.1
Radium production	2	1950–1960	8.8	3.5

Plant No. 12 developed a technology for producing bars of high-grade ^{235}U using natural uranium as a simulator. The ceramics laboratory of NII-9 manufactured the above-mentioned crucibles using a technique and equipment devised with the active participation of A.G. Samoilov and M.A. Dunskiy. Hermetically packed batches of crucibles were brought by air to Plant V of Center No. 817 where components for nuclear charges were produced of metal ^{235}U.

Dangerous technological operations were conducted at Plant No. 12 when fluorine was used in the metallurgical process.

In the early years uranium hexaflouride was produced at plants of the People's Commissariat of Chemical Industry. A.S. Leontichuk, a veteran of the atomic industry, one of the pioneers of producing UF_6 at the Siberian Chemical Center, recalls that the first cylinders with UF_6 were delivered to Plant No. 12 from Dzerzhinsk in 1947. A technology for UF_6 production was devised by NII-42, the head institute of the People's Commissariat of Chemical Industry, according to directions given by M.G. Pervukhin. The first 10 g of UF_6 were produced at the NII-42 laboratory headed by B.A. Alekseev as early as the second half of 1943. In mid-1944 the People's Commissariat of Chemical Industry decided to construct a special workshop (designed by the GSPI-3 institute under the guidance of A.S. Fedulov, P.G. Khain, and B.A. Alekseev, scientific director of the project) at the Rulon Plant in Dzerzhinsk. Implementation of the technology for producing uranium metal from UF_6 devised by the NII-9 called for setting up a new production process at the Rulon Plant. Later, the production of elementary fluorine, UF_4, and UF_6 was arranged at a Kirov-Chepetsk Plant and then at the Siberian Chemical Center [143]. The first designers of the conversion of UF_6, delivered to Plant No. 12 in 1947, to uranium oxides and uranium metal were N.P. Galkin, S.I. Zaitsev, and M.M. Gorin. Apart from UF_6, the product delivered from Dzerzhinsk contained vanadium, chromium, phosphorus, molybdenum, and other unwanted elements. When manufacturing uranium cores to be loaded to the first commercial reactor, the content of impurities was analyzed in each of the refining phases.

On January 28, 1948, a governmental decree was issued which charged four institutes (VIAM, NIIKHIMMASH, NII-9, and NII-13) with devising a method for sealing uranium slugs destined for the first commercial reactor. This decree demanded that each of these institutes should provide the experts' appraisal of methods for coating uranium cores with a protective jacket devised by other institutes [24]. The sealing of a uranium core was to prevent its contact with water, used as a coolant of the reactor core, and ensure the invariability of the slug size. A program for testing uranium slugs was drawn by the LIPAN institute. On August 26, 1946, this program was sent to all of the institutes concerned. V.V. Goncharov [24] notes that they planned to conduct the following tests using the NIIKHIMMASH (Research Institute for Chemical Machine Building) stand:

- shock deformation caused by a fall down in a reactor fuel channel (approximately 20 m);
- canning tightness;
- heat conductivity and abrupt temperature changes;
- swelling of the canning and surface of slugs.

A special commission consisting of V.S. Emeliyanov, V.F. Kalinin, V.I. Merkin, and V.V. Goncharov conducted tests of 5 batches, each including 20 slugs, and declared the VIAM and NII-13 projects to be the best. These institutes were charged with a further research into the sealing of uranium cores.

Researchers from the People's Commissariat of Electric Industry also worked on improving the technology for producing metal uranium. For instance, in 1946 Professor A.S. Zaimovskiy conducted experiments in NII-627 on smelting a rough ingot of uranium in a high-frequency furnace. He recommended to install high-frequency AYaKS furnaces to refine the metal at Plant No. 12. At that time several researchers from academic institutes worked at the plant. One of them was Academician A.A. Bochvar[6] who previously worked for the Institute of Non-Ferrous Metals and Gold and other research institutes. As early as 1936 he developed a method for crystallizing shaped castings under pressure. Under the scientific guidance of A.A. Bochvar, A.S. Zaimovskiy, and A.G. Samoilov [38], Plant No. 12 developed a technology for manufacturing hemispherical pieces from metal uranium of natural enrichment. Hemispheres of various diameters were used at Center No. 817 to determine the critical mass of solutions enclosed in spheres equipped with uranium reflectors of various thicknesses [144]. Later, using the experimental facilities of Plant No. 12, they developed a technology for fabricating other hemispheres of high-grade uranium destined for uranium bombs which were manufactured at Plant V of Center No. 817. In 1946, A.A. Bochvar and A.S. Zaimovskiy[7] were transferred to NII-9.

However, their basic objectives at that time were improving technological processes used at Plant No. 12 for producing articles of metal uranium destined for new commercial reactors, and also experiments with plutonium and ^{235}U at Plant V, Center No. 817.

The basic objective of Plant No. 12 was to master the industrial (large-tonnage) technology for producing uranium slugs. On August 15, 1946, I.V. Kurchatov reported to the Soviet Government that the canning of uranium slugs was permanently given much attention, and that a large body of research was conducted at Plant No. 12 and at the institutes concerned with this problem. The process of canning production and the sealing of uranium slugs consisted of the following operations:

-   stamping of barrels from aluminum sheet bars;
-   forcing the barrels charged with uranium cores through a gauge die;
-   spinning the lid;
-   heating-up the articles placed in the assembly;
-   abrupt cooling of heated assemblies in cold water;
-   etching and oxidizing of the slugs.

Subsequent tests revealed that the technology devised by VIAM and NII-13 could be used for manufacturing uranium slugs for the first production reactor. The NII-13 institute devised a method for applying a nickel coating to the surface of a uranium core which was enclosed in the aluminium canning and sealed by the VIAM technology. Thus, it is commonly agreed that R.S. Ambartsumyan, A.M. Glukhov (VIAM), and P.P. Pytlyak (NII-13) and their teams made a major contribution to the development of a technology for sealing uranium slugs at Plant No. 12. The suitability of uranium slugs for loading in the production reactor was checked also in reactor F-1 at LIPAN.

In the second quarter of 1947, Plant No. 12 completed the production of uranium slugs for the first commercial reactor. Much efforts in implementing this top priority project were made by the plant managers Yu.N. Golovanov, A.N. Kallistov, S.I. Zolotukha, and N.F. Kvaskov.

The production of uranium slugs from metal uranium was permanently subject to improvement, and its output was increasing. New nuclear reactors designed for plutonium production were put into operation first at Centers No. 817 and No. 816 and later at other enterprises. A successful operation of Plant No. 12 was highly appreciated by the Soviet Government. In late 1949 A.N. Kallistov and Yu.N. Golovanov were awarded with the Lenin Orders and Gold Star medals. Several researchers were awarded the State Prizes for the implementation of industrial production of uranium slugs. Improvement of the technology for producing radiation-resistant uranium alloys and manufacturing uranium slugs destined for plutonium production was also a major objective of NII-9. These researches were conducted by the laboratories and teams headed by G.Ya. Sergeev, V.V. Titov, K.A. Borisov, and others under the guidance of A.A. Bochvar and A.S. Zaimovskiy. High-resistant uranium alloys were produced with the participation of these researchers. Workers of the plant and

NII-9 were awarded Lenin Prize for these achievements. (Apart from the workers of Plant No. 12 and VIAM, Lenin Prize was awarded to A.A. Bochvar, G.Ya. Sergeev, and V.V. Titova.

The increasing output of uranium production and the increasing capacity of the reactors, accompanied by the increase of neutron flux density and the growing accumulation of plutonium in the uranium slugs, augmented the amount of research into the production of various uranium alloys. This called for a more comprehensive examination of structural alterations taking place in the crystalline lattice of uranium [145] and for greater amounts of X-Ray diffraction analyses.

One can imagine how arduous and crucial was the work of various services and laboratories of Plant No. 12 in sorting chemicals, raw materials, and substances destined for manufacturing uranium slugs and other products.

Other Materials and Items Manufactured for Nuclear and Thermonuclear Weapons

As soon as Center No. 813 obtained another type of nuclear explosive, enriched ^{235}U [8], Plant No. 12 became the first enterprise where uranium slugs were produced, first with a 2-percent ^{235}U enrichment and later with a higher concentration.

In August 1950 a decision was made to construct the first Soviet reactor designed for the production of special-purpose isotopes. A reactor known as AI with a power output <100 MW was put into operation on December 22, 1951, in Chelyabinsk-40. It was built during a year and a half. This rush was caused by a demand for tritium needed for thermonuclear weapons [113, 146]. Lithium enriched in light isotope ^6Li was the most suitable material for producing tritium. V.S. Emeliyanov, Deputy Director of the PGU, recalls in his book devoted to the memory of Kurchatov that a thermonuclear project called for the creation of production facilities as complicated as in the case of ^{235}U and plutonium [11]. To this end, it was necessary to start a research into the production of heavy hydrogen isotopes (deuterium and tritium) and into the separation of lithium isotopes.

Pure lithium was obtained in 1855 by German chemist R. Bunzen. It consists of ^7Li (92.7%) and ^6Li (7.3%). Because tritium is produced from ^6Li, lithium compounds enriched in ^6Li are used in the atomic industry. For this reason lithium is as important for the production of thermonuclear weapons and power industry, as uranium for nuclear power industry and plutonium production.

In the Soviet Union the problem of creating a hydrogen bomb was first posed in a special memorandum submitted to the Soviet Government in 1946. The authors of this memorandum were researchers from Laboratory No. 2: I.I. Gurevich, Ya.B. Zeldovich, I.Ya. Pomeranchuk, and Yu.B. Khariton [80]. In 1948, at the FIAN institute (Director S.I. Vavilov), a group of theorists was set up headed by world-known scientist I.E. Tamm.[9] A governmental decree charged this group with studying a possibility of creating thermonuclear weapons. By that time it was already known that a research into the creation of a hydrogen bomb was under way in the USA.[10] The US researchers were studying an addition reaction between deuterium nuclei (^{2}H) and a heavier nucleus of hydrogen — tritium (^{3}H).

The synthesis of heavy hydrogen nuclei can take place under extremely high temperature and liberate a considerably higher energy than the energy produced by the fission of uranium or plutonium nuclei. At a temperature of 20 million degrees Celsius no instantaneous release of energy or thermonuclear explosion can be obtained using lithium or deuterium isotopes. In the case of tritium the reaction rate is higher than in the case of deuterium synthesis by a factor of ca. 200. At a higher temperature the rate of the synthesis of lithium nuclei and hydrogen isotopes is sufficient to produce a thermonuclear explosion. This high temperature can only be obtained using an atomic bomb as a fuse. Yu.A. Romanov, a designer of a thermonuclear weapon, a member of Sakharov's team writes [80]: " ...if you simply surround such a fuse with liquid deuterium, this would not give any material increase of the explosion power... Because of the dispersion of fragments at high pressure and a temperature drop due to heat transfer, the synthesis reaction rate is too low, and only a negligible amount of deuterium nuclei have enough time to react." The situation is much better if tritium is used: it reacts with deuterium at a rate hundreds of times higher than in the previous case. However, there is no tritium in nature. To produce tritium by the reaction

$$^6_3\text{Li} + n \rightarrow {}^3_1\text{H} + {}^4_2\text{He},$$

one should have lithium enriched in ^{6}Li and a nuclear reactor producing a high neutron flux.

Thermonuclear weapons can be created using a fission of lithium and deuterium isotopes, but in this case a much higher temperature is needed to initiate this reaction. The use of a shell made of natural uranium and a homogeneous mixture of deuterium and tritium was not the only contribution of A.D. Sakharov to the creation of thermonuclear weapons. He also proposed a heterogenous structure consisting of alternating layers of a light substance (deuterium, tritium, or their compounds) and heavy isotope ^{238}U, which he called "flaky pastry." The idea of using ^{6}Li in a "pastry" belonged to V.L. Ginzburg, a researcher from FIAN.[11] He was the author of the name "Lidochka" given to lithium deuteride (LiD), probably for secrecy [80]. To produce "Lidochka," they had to arrange a number of production processes including the separation of isotopes of light elements. In addition to enterprises of the chemical industry they recruited enterprises of the PGU, including Plant No. 12, to solve these problems.

As soon as the AI reactor was put into operation in late 1951, it was used for tritium production. A technology for the recovery of tritium from irradiated lithium slugs and the production of compounds needed for thermonuclear weapons was imperfect, posed hazards, and was permanently improved. The first test of a thermonuclear weapon was conducted at the Semipalatinsk test site on August 12, 1953.[12] V.A. Malyshev, Minister of the Minsredmash Ministry, chaired the State Commission. Among its members were I.V. Kurchatov, I.E. Tamm, A.D Sakharov, B.L. Vannikov, and Marshal A.M. Vasilevskiy. Academician E.N. Avrorin notes that in fact the first Soviet thermonuclear bomb was tested in November 1955; it was a thermonuclear device that was tested in 1953 [147].

Uranium slugs, enriched in ^{235}U, rather than natural uranium are usually used in reactors to produce large amounts of tritium and other isotopes with a high neutron absorption. The excess neutrons emitted in ^{235}U fission are captured by targets to produce useful isotopes. The targets used to produce tritium are made of slugs enriched in ^6Li. The amount of plutonium produced in the uranium reactor slugs consisting of high-grade ^{235}U sharply drops because the amount of ^{238}U, which converts to plutonium on neutron capture, decreases with the concentration of ^{235}U by nearly 3 times with a 2-percent enrichment. When using uranium slugs with a 90-percent enrichment, the amount of ^{238}U and, consequently, the specific yield of plutonium become many times lower. Moreover, plutonium thus produced and accumulated in uranium slugs is unsuitable for military purposes. At a high burnup, a large amount of useless isotope ^{240}Pu is accumulated in plutonium.

The rate of a thermonuclear reaction is highly affected by the density of a mixture of hydrogen and lithium isotopes. There must be no heavy elements in it because they increase its heat capacity and impair the reaction efficiency [69]. A compound that meets the requirements of high density and absence of inactive impurities is the salt-like compound of hydrogen and lithium, lithium hydride (LiH), whose density under ordinary conditions is 0.82 g/cm^3 (LiH may occur in six isotope combinations: ^6LiH, ^7LiH, $^6Li^2H$, $^7Li^2H$, $^6Li^3H$, and $^7Li^3H$).

The problems related to the creation of thermonuclear weapons are beyond the scope of this book. Moreover, they were discussed in detail by Yu.A. Romanov [80] and also by V.I. Ritus in his sketches of Sakharov's scientific portrait [148]. The contributions of I.E. Tamm, A.D. Sakharov, V.L. Ginzburg, and other scientists to the creation of thermonuclear weapons is well known. What is worth mentioning here is the fact that the source materials used in manufacturing these weapons were produced at Plant No. 12 and at other enterprises before it was built.

The major contribution to the development of a technology for separating isotopes of light elements was made by Academician B.P. Konstantinov.[13] Under his supervision a method for isotope separation was devised, which was used as a basis for organizing a large-scale commercial production [5].

Upon completion of its reconstruction, Plant No. 12 proceeded with the production of ^{6}Li and its hydrides. Usually, small old workshops where conventional munitions had been produced were used for the production of new materials.

By that time, in addition to facilities producing ^{6}Li and its hydrides, there were workshops supplying special filters and magnets for the plant equipment applied in concentrating the uranium isotopes (Table 22). Since 1950 the plant was also recovering radium from uranium for various consumers. Besides, the plant produced radiation sources, including the radium ones.

The plant was permanently reconstructed because some of the mastered technological production processes were transferred to other enterprises. Until 1954, chemical and metallurgical production processes predominated at the plant. When the first NPP was put into operation, the plant proceeded with devising technologies for manufacturing fuel elements for NPPs, navy nuclear power units, and all the types of experimental reactors constructed in the Soviet Union and East European states.

Table 22 Basic products of Plant No. 12.

Product	Years	Production area, thousand m^2	Investment, mln roubles
Diffusion filters	1948–1960	4.3	1.5
Neutron and gamma-ray sources	1956–1967	2.1	1.1
Lithium	1956–1962	7.2	10.2
Lithium metal and lithium hydrides	1956	4.0	1.9
Magnets	1960–1967	5.5	1.9

In contrast to commercial nuclear reactors, fuel elements of the nuclear power plants remain in the reactor core at least ten times longer, 2 to 3 years. Their reliable operation must be ensured throughout this period of time. In the early years there were many problems.

In 1948–1949 the effects of the radiation growth of uranium and graphite were discovered in the first commercial reactor A. In order to study the mechanism of these effects it was necessary to arrange a special facility at the LIPAN consisting of an experimental nuclear reactor (RFT) and a laboratory for hot materials technology destined for handling highly radioactive substances. A proposal for building them was submitted by I.V. Kurchatov and A.P. Aleksandrov in 1949. Moreover, they proposed " ...to build another reactor of the RFT type, with a higher power output, at Plant No. 12 in Elektrostal City to do research into materials technology because of a too large quantity of fuel elements that were to be tested at the LIPAN RFT reactor" [24, 99].[14]

This research center was constructed at LIPAN in 1952. Plant No. 12 had no reactor of this type, and the fuel elements produced at the plant for all research reactors, NPPs and navy reactors were tested at LIPAN.

The nature of uranium radiation growth under various irradiation conditions was examined under the scientific supervision of S.T. Konobeevskiy. The improvement of a technology used at Plant No. 12 to produce more reliable fuel slugs, resistant to uranium radiation growth was implemented by researchers at NII-9 under the supervision of A.S. Zaimovskiy. An important contribution was made by scientists from other institutes to the solution of this problem. For instance, a technology for manufacturing dispersion fuel elements for the first NPP was developed by a team of researchers from the FEI institute headed by V.A. Malykh.

Experimental reactors RFT (LIPAN), AI (Center No. 817), and TVR (Laboratory No. 3) could not be devised without using fuel elements of enriched uranium. Fuel elements used in the first experimental nuclear reactors had uranium with a 2-percent ^{235}U enrichment. Later, ^{235}U enrichment in most of the experimental reactors was raised first to 10–36% and then to 90%. The constructions of fuel elements are displayed in Fig. 27. Contrary to the simple uranium slugs used in the uranium-graphite reactors (including reactors of the Siberian NPP), the uranium slugs employed in experimental reactors can be circular, multilayer of various shapes, where each of the uranium-containing layers is sealed into a jacket made of aluminum, steel, zirconium, or various alloys. Bundles of such fuel elements are called "fuel assemblies." Each assembly contains nuclear fuel of enriched uranium (plutonium) which is accommodated in the matrix consisting of a nonfissile substance with a low neutron capture section. These are known as dispersion fuel elements. By the radiation stability of a fuel element is meant the capability of keeping its size invariable throughout the entire irradiation cycle without inducing any disturbances in the reactor operation. The fuel element must hold all fission products throughout its operation period, and its jacket must remain leak-tight. The radiation stability of fuel elements depends on the following factors:

- grain size and volume of a fissile phase and the structure of a dispersion composition;
- type of the fissile compound used as a fissile phase in the dispersion composition;
- properties of the matrix materials and the effect of a manufacture technology on them;
- conditions of fuel element operation in a reactor (temperature, thermal cycles, vibration, cavitation, cooling, etc.);
- variations in the properties of fission material and matrix on irradiation;
- design of the fuel element and jacket material;
- type of the coolant and presence of impurities in it that may cause corrosion cracking;
- presence of precipitate on the surface of a fuel element.

Because of the high complexity of the problem, its solution was possible only on condition of the joint efforts of the personnel of NPPs, scientific supervisors, chief designers of reactors, designers of the fuel elements (NII-9, VIAM, and other institutes), and technologists from Plant No. 12.

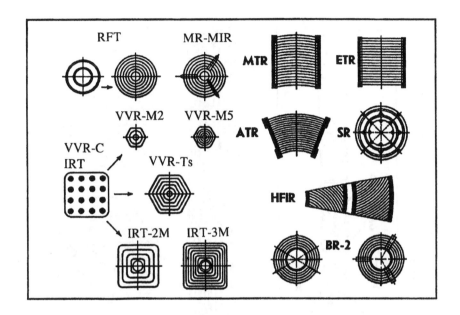

Fig. 27 Sectional view of fuel assemblies used in Soviet (left) and Western (right) reactors.

A comprehensive description of the bulk of dispersion fuel elements employed in various reactors and numerous data on the manufacture of these elements at Plant No. 12 are presented in [60, 149] and in Fig.27.

Dioxide of enriched uranium (UO_2) is used as fuel in most of NPP reactors (VVER and RBMK).

Compared to uranium metal, UO_2 has a much higher radiation stability. Comparing the growth of UO_2 volume and that of uranium metal, Goncharov [24] found that the UO_2 growth rate was lower by a factor of 12. This conclusion was based on numerous studies conducted at the LIPAN (V.V. Goncharov, P.A. Platonov, A.D. Amaev, and others), at NII-9, and at other enterprises under the scientific supervision of S.T. Konobeevskiy. All experimental batches of fuel elements and assemblies destined for reactor experiments were manufactured by the workers of Plant No. 12. This plant and NII-9 institute jointly developed a technology for producing uranium oxide and other uranium compounds.

Best studied was the radiation stability of UO_2 fuel elements which found a wide application in the nuclear power industry.

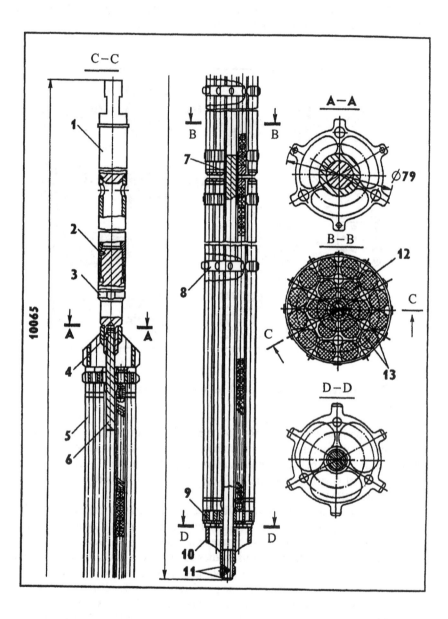

Fig. 28 Fuel cassette for an RBMK reactor: 1 – hanger; 2 – pin; 3 – adapter; 4 – liner; 5 – fuel element; 6 – bearing rod; 7 – bush; 8 – support plate; 9 – limit plate; 10 – fitting; 11 – nut; 12 – frame tube; 13 – support plate cell.

Fuel compositions, elements, and assemblies for the first NPPs were manufactured at Plant No. 12. The fuel assembly of the RBMK reactor comprises 18 fuel elements, whereas in the VVER reactor the number of such elements is many times greater. A fuel assemblage complete with all elements providing loading and unloading of nuclear fuel is called in Russian *cassette*. Figure 28 shows a cassette of a serial RBMK reactor, comprising 13 basic pieces and units. Some of them have a complex structure and are manufactured of special steel, zirconium, and other materials.[15] The length of cassette consisting of two fuel assemblages is >2 m. The number of fuel elements loaded into one RBMK reactor is ca. 60 thousands. A fuel element contains fuel pellets of baked dioxide; its jacket is made of an alloy of zirconium and 1% of niobium.

Plant No. 12 had all facilities necessary for the production of fuel itself, fuel elements, and fuel assemblies: chemical-metallurgical facilities, assembly shops, various test benches, and working sites for cutting fuel elements and dissolving fuel compound if necessary.

Specific requirements to the sizes and quantity of fuel elements and fuel assemblies and an increasing demand for these articles in the nuclear power industry that evolved since 1954, as well as a necessity for improving the working conditions for the personnel, made it necessary to construct automated workshops and working sites and reconstruct the plant. During the first 20 years, 96,000 m^2 of new production areas were added, and more than 240,000 m^2 of the existing production areas were reconstructed. These operations were implemented by many design, construction, and erection-building organizations.

The successful work on manufacturing fuel elements at the plant was highly appreciated. S.I. Zolotukha, T.A. Purtova, A.N. Partin, I.T. Obraztsov, D.D. Sokolov, B.I. Ermolov, Z.V. Sadchikova, and I.I. Kuznetsov became Lenin Prize winners. Many people got other governmental awards.

Up to now, Plant No. 12 remains the main facility that provides reactor fuel elements for all reactor research centers of Russia and the states of the former USSR [98]. Fuel elements and assemblies for RMBK-type reactors and navy nuclear power units are still the main products of Plant No. 12.

NOTES

1 Later he became a member of the VGU Board, worked as Director of the Scientific and Technical Department of the Minsredmash Ministry (Medium Machine Building Ministry), and as Director of the VNIIKhT institute; Corresponding member of the Academy since 1946.

2 Subsequently, Laboratory No. 2 was called LIPAN, and after the death of Kurchatov it was called Kurchatov Institute for Atomic Energy (IAE).

3 A prominent specialist and manager in production of rare metals and atomic

industry. In 1938–1945 he worked as a chief engineer and director of the Lead-Zinc Directorate, People's Commissariat of Nonferrous Metals. From 1945 he worked for the PGU first as a chief engineer and then as a deputy director of the 2nd Directorate. In 1949–1953 he worked as a director of the 1st Directorate of the PGU, and in 1953–1974 as a director of the 3rd Chief Directorate of the Minsredmash Ministry.

4   Later, this facility and a few laboratories of NII-9 dealing with the recovery and processing of uranium ore were transferred to NII-10, where A.P. Zefirov and, subsequently, D.I. Skorovarov worked as directors.

5   In 1947–1948 the Analytical Council concentrated its efforts on the analysis of products of Plants B and V, Center No. 817.

6   In the early years A.A. Bochvar headed an experimental workshop at Plant No. 12 (Building 80).

7   Subsequently, he worked as Deputy Director of the VNIINM institute, winner of Lenin and State prizes, took an active part in devising technological processes for producing uranium slugs at the Minsredmash plants.

8   It was impossible to produce enriched ^{235}U (especially in the early period) without the diffusion filters manufactured at Plant No. 12.

9   P.A. Cherenkov, I.E. Tamm, and I.M. Frank were awarded a Nobel Prize in physics for the discovery, interpretation, and utilization of the Cherenkov effect.

10 E. Teller was a pioneer of creating a hydrogen bomb in the USA (1942).

11 He worked together with A.D. Sakharov in Tamm's team; Corresponding Member of the Academy since 1953; Academician since 1966.

12 Immediately after the explosion V.A. Malyshev, A.D. Sakharov, I.V. Kurchatov, and some other researchers inspected the test epicenter and received high radiation doses [31].

13 He worked for the LFEI institute, first as the head of a laboratory, then of a department, and then as director; Corresponding Member of the Academy since 1953; Academician since 1960; Vice-President of the Academy since 1967.

14 Reactor RFT, consisting of 5 annular units, was the first world's materials technology reactor of the annular type providing a record neutron flux density of 5×10^{13} neutron/cm$^2 \cdot$ s.

15 Zirconium alloys and zirconium fuel channels of the RBMK reactors, as well as the outer tubes of fuel elements (made of special alloys), are manufactured at a plant in Glazov City.

CHAPTER 12

Nuclear and Radiation Safety

The creation of independent armed forces and independent energy policies of the CIS states (former Soviet Republics), as well as the necessity of ratifying Russia's agreements assuming the reduction of strategic offensive armament in the nuclear states, created complex problems arising from handling the accumulated reserves of plutonium and highly enriched uranium applied in manufacturing nuclear weapons. These materials, as well as fuel elements used in various nuclear reactors (especially in experimental reactors that operate in the CIS states) are a potential source of high hazard. These problems have a long-term, mostly tragic history. Particular accidents which occurred in handling fissile materials were described in memoirs by Minister E.P. Slavskiy, Academician A.P. Aleksandrov, and V.N. Chernyavin, Admiral of the Fleet [116].

Information about the inspection of nuclear weapons and violations in handling nuclear ammunition was brought to the public notice. Managers of Russia's Federal Nuclear Centers V.A. Belugin (Arzamas-16) and V.Z. Nechai (Chelyabinsk-70), researchers from these centers, and chief designers of nuclear charges and ammunition A.N. Senkin, A.A. Brish, S.N. Voronin, G.N. Dmitriev, and B.V. Litvinov discussed these problems in their publication [150]. They emphasized that accidents caused by violations in the storage and handling of nuclear ammunition might result in effects comparable with the Chernobyl catastrophe.

Particular aspects of the nuclear safety problem are discussed below without analyzing the attitudes of other states possessing nuclear weapons and fissile materials. As known from literature more than 20,000 nuclear warheads comprising tens of tons of plutonium and a few tons of high-concentrated ^{235}U

are located in the CIS area. The bulk of these materials are concentrated in strategic offensive armaments (Table 23).

In addition to the materials listed in Table 23, fissile materials produced from enriched ^{235}U are contained in the fuel elements of various vessels equipped with the nuclear power units. Plants producing enriched ^{235}U, fuel elements, and assemblies, and also radiochemical plants are located on the territory of Russia. Nuclear fuel pellets made of UO_2 are produced in Kazakhstan (Ust-Kamenogorsk). Highly-enriched ^{235}U is used in reactors located in Latvia, Georgia, and Armenia. Table 23 does not give complete data on all fissile materials or the number of warheads contained in the strategic nuclear weapons situated on the territory of the above-mentioned states. The amounts of uranium and plutonium contained in the nuclear fuel unloaded from the reactors are not included either. In 1992, 2,605 tactical nuclear warheads were located in the territory of the Ukraine [151]. In all of the above-mentioned experimental nuclear reactors, fuel rods mostly contain ^{235}U with up to 90-percent enrichment.

Table 23 Distribution of basic fissile materials in the equipment located in the CIS states and Lithuania [151, 152].

State	Strategic offensive armament		NPP, GW	Research reactors
	Number of types	Number of warheads		
Russia	4	7449	21.3	25
Ukraine	2	1408	12.9	1
Kazakhstan	2	1360	0.15	2
Belarus	1	54	–	1
Lithuania	–	–	2.5	–

^{235}U and ^{239}Pu are also contained in various critical assemblies and subcritical systems. The inspection of more than two hundred of nuclear and radiation-hazardous enterprises of Russia conducted in late 1993 revealed the following facts:

– the growing amount of spent nuclear fuel stored at the NPPs impairs nuclear safety;
– nuclear power units of the naval and merchant vessels are often operated with departures from the nuclear safety regulations now in use.

In the concluding report it was noted that the State Atomic Inspection of Russia keeps records of 116 research reactors, critical assemblies and subcritical systems. The main conclusion based on the inspection of the nuclear enterprises belonging to Ministry of Atomic Industry, Russia's Defense Committee, Defense Ministry, Russia's Research Center "Kurchatov Institute" and other departments was as follows [153]: "The state of nuclear and radiation safety in the Russian Federation is unsatisfactory."

The nuclear fuel used in NPP reactors (VVER-440, VVER-1000, RMBK) contains 1.8-4.6% ^{235}U. From the nuclear safety standpoint handling of this fuel poses a substantially lower hazard. However, under certain conditions self-sustaining chain nuclear reactions may occur. The large research and production centers of Russia gained a great experience and solved many problems in providing nuclear and radiation safety when handling fissile materials. The fact that materials contained in the nuclear weapons are of great value makes it clear why the Ukraine wants to use them, in particular, as a nuclear fuel. However, as stated by the managers of the Russian Federal nuclear centers and the designers of the Soviet nuclear weapons, only irresponsible politicians can propose the creation of Ukrainian centers for the disassembling and dismantling of the nuclear weapons left there: "None of professionals recognizing his responsibility for our planet will set about disassembling nuclear armaments developed in another country by an unknown technology."

It is well known that the Ukraine did not give up the concept of its own nuclear weapons and even received support among a few states. The Canadian Minister of Foreign Affairs noted at a session of 53 states — participants of the European States Safety Council [154] that Canada believed Kiev's statement that Russia could not firmly guarantee the demolition of nuclear weapons taken away from the Ukraine. Canada proposed its intermediary assistance in negotiations to promote the solution of this problem. It should be noted that from the standpoint of the designers of nuclear armaments, demolition means the dismantling of ammunition and the demolition of its basic components with the exception of plutonium and ^{235}U. The rest of the fissile materials cannot be considered as ammunition and can be stored or used as fuel in thermal- and fast-neutron reactors.

There were many dramatic instances in the history of mastering atomic energy in Russia and in other nuclear states. Some of them were caused by a unique ability of nuclear materials to sustain a spontaneous chain nuclear reaction accompanied by an intense emission of radiation and a release of radionuclides. These reactions may occur in materials occurring in any phase state — solid, liquid, or gaseous. The basic condition for the initiation of a spontaneous chain reaction is the amount of fissile material inside a system sufficient for the formation of a critical mass M_{cr}. The safe amount of fissile material depends on its concentration in a solution (gas or solid), and also on the shape and size of the enclosing vessel. For these reasons the notions of "nuclear safe geometry of equipment" and "nuclear safe concentration of fissile materials" have been introduced in the atomic industry. A spontaneous chain reaction starts when the number of neutrons in a particular system grows in an avalanche manner (instantaneously), i.e., when the effective neutron multiplication factor (K_{eff}) in a system containing fissile materials is ≥ 1. To produce the explosion of a nuclear device one should provide conditions where $K_{eff} > 1$. In nuclear reactors one should maintain conditions where K_{eff} is approximately equal to 1.

The critical masses of ^{235}U, ^{239}Pu, and ^{233}U were determined in the early years of the creation of nuclear weapons. By the critical state of a system (state of a self-sustaining nuclear chain reaction) is meant a state of a system where the number of neutrons remains invariable. This can be provided only if the neutron absorption and leakage are equal to the number of neutrons produced in the nuclear reactor by the fission of ^{235}U, ^{239}Pu, or ^{233}U nuclei. This ensures the condition of $K_{eff} = 1$. One can identify this state experimentally using either of the following two means:

- critical experiments where conditions are provided to make $K_{eff} \geq 1$;
- subcritical experiments where the neutron flux is precisely measured at the test table (subcritical assembly) as a function of loaded fissile materials. In this case only an approximate value of the critical mass is obtained.

In the second case $K_{eff} < 1$, and the accuracy of critical mass determination depends on the proximity of the experiment to $K_{eff} = 1$.

Table 24 Critical masses of the first research reactors fueled with natural uranium [105, 112].

Reactor state	Year of start	M_{cr}. tons	Moderator, tons	Power output	Average thermal neutron flux, $cm^{-2} \cdot s^{-1}$
SR-1, USA	1942	46	Graphite, 385	100 W	4×10^6
SR-2, USA	1943	52	Graphite, 472	2 kW	1×10^8
SR-3, USA	1944	3	Heavy water, 6.5	300 kW	5×10^{11}
X-10, USA	1943	35	Graphite, 620	3800 kW	5×10^{11}
F-1, USSR	1946	45.1	Graphite, 450	1000 kW*	4×10^7
HWR, USSR	1949	1 25**	Heavy water. 4.18***	max 500kW	2×10^{12}

* The power output was as high as 3860 kW at high-power start-ups.
** Diameter and length of uranium slugs were 2.2 cm and 162.5 cm, respectively, with a square grid spacing of 12.7 cm. The critical D_2O level was 174.5 cm.
*** The reactor had a graphite reflector enclosing the outer and side faces of the aluminum tank (175 cm in diameter, the thickness of the walls and bottom was 3-3 2 mm).

In the early years, both in the USA and in the USSR a critical mass was determined only in experiments with natural uranium. In these experiments they first used graphite and then heavy water as a moderator (Table 24). In the USSR, the first critical experiments in the uranium–heavy water system were conducted in 1949 at Laboratory No.3 under the guidance of Academician A.I. Alikhanov.

Heavy-water reactors have a number of specific features as compared to graphite reactors:

- relatively small size of the system, which results in a high leakage of neutrons;
- small number of collisions necessary to slow down neutrons;
- production of secondary neutrons through reaction $D(\gamma, n)H$;
- high cost of heavy water.

The knowledge of the minimum critical mass values of the uranium–D_2O system was a critical requirement in creating not only an advanced experimental heavy-water reactor but also production heavy-water reactors which were later constructed at Center No. 817 for producing first plutonium and then ^{233}U and tritium isotopes.

General Safety Problems in the Early Years of Atomic Industry

Development of atomic industry first in the USA and later in the USSR was initiated by the creation of nuclear reactors comparable with the largest hydroelectric power stations in power output. The only difference was that contrary to the latter, the first production reactors were constructed for producing plutonium, and that all energy they produced was spent for heating the Columbia River in the USA and the Kyzyl-Tyash Lake in the USSR. Admitting the fact that uranium reactor was the greatest achievement of the human mind throughout the history of mankind, F. Soddy, the distinguished English radiochemist, wrote (translated from Russian) [2]: "It is an awful thought that science has prematurely placed the power which seemed absolutely inaccessible to us slightly more than four years ago at the disposal of unprepared men."

The first enterprises dealing with the processing of uranium irradiated in the nuclear reactor and plutonium extracted from this uranium (Plants B and V) faced serious problems of nuclear and radiation safety. For lack of plutonium no experimental data were available on its critical masses both in solution and in other states. Because of this, plutonium masses loaded into technological units were rated with an excess safety factor on the bases of approximate calculations. However, sometimes these high safety factors turned out to be hypothetical and did not prevent the initiation of a self-sustaining chain reaction.

Somewhat different problems arise in the production of highly enriched ^{235}U.

Apart from plutonium, other radionuclides with an activity of hundreds of million curies are produced in commercial and experimental nuclear reactors. So, nuclear and radiation safety should be provided for both the personnel of these facilities and the population. As a matter of fact, there is no problem of radiation safety at the enterprises producing highly enriched uranium.

Great hazards are posed by situations resulting in uncontrollable acceleration of a nuclear reactor of any type. Catastrophic consequences may affect an adjacent area as well as a larger region. Of substantially lower danger are cases dealing with violations in the nuclear and radiation safety rules at research centers where experimental nuclear reactors, pilot critical assemblies, accelerators of charged particles, radiochemical laboratories and other facilities comprising radiation sources are located. No nuclear hazard exists at enterprises dealing only with radioactive sources, except that radiation safety should be provided.

Nuclear and radiation accidents that took place throughout the world in 1944–1988 and were accompanied by the removals of radionuclides yielded the following statistics (accidents that took place in the USSR are not included):

- occurrence of spontaneous critical masses in equipment and chemical processes – 18;
- escape of radionuclides from sealed radiation sources, including accelerators of charged particles and other facilities – 219;
- leakage of radioactive isotopes, including actinides, tritium, and uranium and plutonium fission products – 59.

In most of those cases there were victims and even lost lives [155, 156]. Table 25 presents data on the number of victims who suffered from violations of the nuclear and radiation safety regulations, from accidents in the Southern Ural (September 29, 1957), Chernobyl NPP (April 26, 1986), and from an accident in Brazil where ^{137}Cs powder kept in a special encapsulated radioactive source was dispersed over a large area of Goiania City.

Table 25 Effects of nuclear and radiation accidents [155, 156].

Number of accidents (location)	Number of victims	Number of severely exposed	Number of individuals lost on site
294 (in the world)	1371	633	37
1 (Chernobyl NPP)	135000	24200*	28
1 (Center No. 817)	14100**	6000**	–
1 (Brazil)	244	20	4
Total 297	150715	25733	69

* Including 237 men who received radiation doses above 100 rem.
** Victims who received radiation doses above 44 rem; 6000 received radiation doses above 100 rem, including residents evacuated from the areas where contamination level was above 3,3 Ci/km².

In the early years of the Soviet atomic industry the nuclear and radiation safety could not be appropriately provided. As mentioned above, uranium slugs at the production reactor failed for a great number of reasons. Technological processes used at the plants dealing with fissile materials were also imperfect. That is why plants and workshops at all atomic enterprises were launched and operated in the first years by experienced specialists from research institutes and design organizations. Many of them were transferred to and got a permanent job at workshops and laboratories of atomic enterprises by their own choice.

The personnel of the plants and research institutes worked with great enthusiasm at that time. Everything was done and made for the first time and on one's own. The safety of servicing personnel was basically provided by a severe technological and administrative discipline maintained since the war time and supported by scientific directors of the projects, chief designers, technologists, and managers of all ranks. However, people worked not only because of their conscientious attitude but also for fear. A severe operational regime was set up, cases of dismissals and sometimes prosecution were rather common. Nevertheless, for lack of particular data, especially on the critical masses of uranium and plutonium solutions and other compounds, the operation of technologists at some of the plants posed a considerable risk. Overcautiousness in setting up technological loading rates for fissile materials caused by the lack of many of experimental data was inevitable and in large part justified, though could not always guarantee nuclear safety.

First Nuclear Emergencies in Dealing with Fissile Materials

It is known that a first uncontrollable nuclear chain reaction occurred in the USA in 1945, several months before the explosion of a plutonium bomb at the Amarillo test site on July 16, 1945. The instantaneous flash of a chain reaction (without any destruction and explosion effects) occurred at a stand in Los Alamos during a study of the effectiveness of a neutron reflector made of tungsten carbide placed in two plutonium hemispheres and covered with a 0.13-millimeter nickel coating, with a total weight of 6.2 kg. An experimenter was manually laying the reflector blocks made of tungsten carbide around the plutonium sphere (15.7 g/cm^3 in density). The addition of a next block initiated a self-sustaining chain reaction; 10^{16} fissions of plutonium nuclei took place instantaneously. To stop the self-sustaining reaction, the experimenter smashed the assembly by hand. He received a dose of 800 R and died 28 days later [64]. Another emergency occurred in Los Alamos in 1946 when they were checking the efficiency of a beryllium reflector using the same plutonium hemispheres. When adjusting a clearance between two beryllium hemispheres, a physicist was unlucky to bring them in contact. As a result a critical mass was formed and a neutron burst was produced. The energy of a self-sustaining chain reaction equivalent to 3×10^{15} fissions was liberated, and the radiation dose was as high as 900 R. The researcher died 9 days later. In 1958, an accident took place in Los Alamos during the cleaning and discharging plutonium from the scrap. Plutonium with a concentration ≤ 0.1 g/l was stored in a 196-liter vessel. It was poured by mistake into another tank, 96.6 cm in diameter and 850 l in volume, which contained nearly 300 l of water-organic plutonium emulsion. The addition of 196 l of a solution containing nitric acid resulted in the separation of phases. The upper layer of the solution, contained 3.27 kg of plutonium in a volume of about 330 l. Yet, a critical mass was not formed by that time. However, as soon as a mixer was started, the lower layer of the solution was pushed against the tank wall, whereas the upper layer occupied the freed volume and acquired a shape close to a spherical one; this caused a self-sustaining chain reaction, during which 1.5×10^{17} fissions of plutonium nuclei took place. Three men were near this tank and were severely exposed: an operator received 12,000 R and died; the two others got 134 and 12 R.

Dubovskiy [64] analyzed 19 accidents, 17 of which took place in the USA. In addition to Los Alamos, accidents took place in Oak Ridge, Idaho, and Hanford. Two accidents occurred at experimental reactors, one in Canada (1952) and one in Yugoslavia (1958). One of them occurred in a heavy-water reactor which was being loaded with natural uranium slugs. An error made while watching the level of the D_2O liquid filling the tank caused a self-sustaining chain reaction. In this accident an energy of 80 MW · s was liberated, which was equivalent to 2.6×10^{18} fissions. Six who were nearby received the doses of 400, 700, 850, 850, and 1100 rem. One of them soon died.

In the USSR the first controllable nuclear chain reaction was initiated at 18.00 on December 25, 1946, in a critical uranium-graphite assembly at Laboratory No. 2. The critical mass was 45.1 tons of natural high-grade uranium, mostly in the form of 35-millimeter slugs being loaded in graphite cells at a spacing of 200 mm. In April 1949, Laboratory No. 3 conducted a controllable nuclear reaction in a uranium–D_2O system with a critical mass $M_{c1}^{min} \approx 1.2$ tons. A chain reaction (with prompt neutrons) was produced in the explosion of a plutonium bomb early in the morning of August 29, 1949.

In the early years, the admissible limits of loaded fissile materials were rated at Plants B and V of Center No. 817 proceeding from the theoretical values of critical masses of plutonium that were found at the LIPAN and were personally prescribed by I.V. Kurchatov. Adherence to these specified limits in order to prevent the occurrence of a self-sustaining chain reaction and prevent personnel exposure was to be supervised by directors and chief engineers of Plants B and V, and also by the scientific directors of these plants.

The first nuclear accident took place at Plant B, Center No. 817, four years after it was put into operation. A spontaneous chain reaction occurred on March 15, 1953, in Department No. 26 where the final product of radiochemical plant B was prepared for delivery to Plant V [157]. Among those who suffered was A.A. Karatygin, Director of the Planning and Production Department of the plant. On that Sunday it was his turn to be on duty. He was to prepare tanks for the acceptance of the finished products, plutonium solution at Department No. 26. This is how he describes the situation just before the accident [157]: "A few batches of finished product were to be manufactured during that night. When I saw that we were short of tanks, I found an extra tank nearby. I called G. Akulova, Director of the Department duty shift, to help me. I was well acquainted with all necessary manipulations. I have completed the operation and cleared receiving tanks for the night batches. As soon as I removed a hose, a blue cold flame appeared, and as if an electric current passed through my body. The content of the tank seethed, and a whistling vapor started gushing out of outlet connection." Having understood that a spontaneous chain reaction began, Karatygin did not leave the accident site but took measures to pump a portion of the plutonium solution over to another tank, and in so doing he stopped the nuclear chain reaction. This accident took place because one of the tanks (the one he was unlucky to use) contained some unrecorded product (presumably about 650 g of plutonium). According to a subsequent estimate, 2×10^{17} fissions took place in the course of this spontaneous chain reaction. Karatygin received a radiation dose of 1000 R, and G. Akulova about 100 R. Karatygin survived; he was ill for a very long time and, finally, he lost his legs.

In his memoirs written in 1974 [157], Karatygin describes in detail the course of his sickness, how he was treated in Hospital No. 2 in Chelyabinsk-40 and Moscow Hospital No. 6 of the USSR Ministry of Health. After undergoing a course of treatment, he worked as a technical editor at the Scientific and Engineering Information Department and at the Central Laboratory of the center.

Reasons for accumulating unrecorded fissile materials at the plants were extremely confidential conditions and a race for a scheduled output of final products. It seems rather absurd nowadays that in the early years the plutonium production plan was not known even to M.V. Gladyshev, Deputy Chief Engineer, and Professor A.P. Ratner, Scientific Supervisor of the plant, who were responsible for technological processes, plutonium loss and recovery, and accident prevention at the plant [43]. Moreover, the accuracy of analytical and remote methods applied to control radiochemical production was insufficient.

The first accident at the chemico-metallurgical production unit of plant V took place on April 21, 1957. In a chamber in workshop No. 1 a spontaneous chain reaction occurred in a horizontal montejus subsequent to delivery of a small amount of 90-percent ^{235}U with a nuclear safe concentration (several grams per liter). An immediate reason for this accident was the accumulation of a precipitate containing 3.4 kg of 90-percent ^{235}U in a horizontal montejus. The form and inferred arrangement of the precipitate and the added solution was sufficient to initiate a spontaneous chain reaction. The analysis of the causes of this accident disclosed that a slow precipitation of uranium took place in the montejus ends where the speed of product-bearing solution motion was lower.

In the presence of accumulated precipitate containing ^{235}U, addition of solution even with an admissible concentration of fissile materials into the montejus caused a spontaneous chain reaction. The energy equivalent to 10^{17} fissions of ^{235}U nuclei was liberated. Six workers suffered. The operator died within 12 days, the others got over a radiation sickness. After this accident, time intervals between the flushings of the workshop equipment were reduced down to 10–15 days. It was clear that one should strictly control the efficiency of cleaning the equipment from fissile materials. To provide a control of nuclear safety at plants B and V, new posts, Deputy Chief Engineers in Science, were established in addition to scientific directors. In the early years M.I. Ermolaev (plant B) and M.A. Bazhenov (plant V) took those positions.

In his paper [158] V. Frolov notes that from 1953 to 1987 thirteen nuclear accidents, including two accidents at critical stands (Center No. 817 and Plant No. 12), took place at three production facilities (Centers No. 816 and No.817 and Plant No. 12). Apart from these accidents, single spontaneous chain reactions occurred in special-purpose physical assemblies and stands at the head research institutes (Kurchatov Institute and FEI).

The basic requirement providing nuclear safety and preventing spontaneous chain reactions is the control of the rated masses of loaded fissile materials, which demands the knowledge of the minimum values of critical masses admissible in the technological equipment of plants and storage facilities, transportation tanks, and active zones of nuclear reactors of all types.

Minimum Critical Masses and Dimensions of Safe Nuclear Installations

In all technological processes that deal with fissile materials there is a potential danger of a spontaneous chain reaction. These reactions, especially when they arise in the equipment of the technological lines of production enterprises, result in inadmissible personnel exposure. They are commonly accompanied by dramatic effects particularly because of their unexpected character. In the early years, spontaneous chain reactions occurred due to violations of the technological processes, mostly because of insufficient knowledge.

No experimental data on the critical mass of plutonium solutions M_{cr} were available when the first Soviet radiochemical and chemico-metallurgical plants were put into operation at Center No. 817. Theoretical M_{cr} calculations made for numerous technological vessels containing plutonium solutions were too rough. Moreover, the situation was aggravated by the fact that the material of technological vessels, their sizes and shapes, as well as the actual concentrations of plutonium in solutions might change the M_{cr} value by a factor of 10. The limiting sizes of nuclear-safe equipment and maximum admissible concentration of plutonium in solutions that excluded the formation of a critical mass in any technological volume (i.e., prohibited a spontaneous chain reaction), were also unknown at that time. That is why some portions of plutonium produced at the industrial reactors (after the delivery of plutonium in amounts necessary for the solution of the major problem – manufacturing of atomic bombs) were used at special facilities and stands for determining the critical masses of various plutonium compounds occurring as solutions, solids, and pure metal.

No open publications about Soviet experiments dealing with the problem of nuclear safety and M_{cr} determination under various conditions were allowed until 1955. However, as was apparent later, the very first experiments with ^{235}U and

[239]Pu solutions were conducted at the Plutonium Center in Chelyabinsk-40 as early as
1950-1951. By then, both plutonium and
[235]U were produced in sufficient quantities to
start a research into the determination of
critical masses. Experiments were made in a
specially constructed stand building at a
distance of about 1 km from the production
reactor A, in a locality known as *Berezki*,
near the site where Facility No. 40 consisting of an experimental design bureau and an
instrumentation plant was constructed later.
All experiments were guided by Kurchatov
who resided in Chelyabinsk-40 at that time.
B.G. Dubovskiy [11] and G.N. Flerov [144]
recalled that in addition to themselves,
among those who participated in the experiments were Ya.B. Zeldovich, E.A. Doilnitsyn, N.A. Dmitriev, E.D. Vorobiev,
I.E. Kutikov, and S.M. Polikanov. They also
recruited workers of the plant for servicing
stand operations.

**GEORGIY NIKOLAEVICH
FLEROV,**
Academician from 1968.
**Discovered spontaneous fission of
uranium nuclei (together with
Petrzhak). Contributed to the
nuclear weapons project.**

Their major objective was to determine
the critical masses of [235]U and [239]Pu solutions in spherical tanks. They made use of
[235]U of a 75-percent enrichment occurring as
a $UO_2(NO_3)_2$ solution with concentrations of 92.2, 184.3, and 368.7 g/l. In
[239]Pu experiments they used $Pu(NO_3)_42HNO_3$ solution with concentrations of
38.4 and 316 g/l. The diameter of spherical vessels made of aluminium and
stainless steel was varied from 150 to 300 mm. For each of these spheres,
spherical neutron reflectors of varying thickness and outer diameter of 500 mm
were manufactured from paraffin, graphite, aluminum, iron, lead, or natural
uranium. In the center of a sphere they placed a neutron source; the growth of
neutron multiplication, as the sphere was filled with the solution, was controlled
using a special start-up equipment. Starting with flasks, spheres of a small
diameter, and successively passing to large-size spheres, they managed to assess
M_{cr} of a spherical volume of solution for each concentration using different
neutron reflectors. Table 26 presents the resulting M_{cr} values obtained in these
first Soviet experiments.

The most important result of these experiments was the determination of a
maximum admissible mass of plutonium in a solution (510 g). This estimate
provided a reliable basis for setting up the admissible limits for plutonium
masses to be loaded into the equipment. Yet, to provide the nuclear safety and

effective utilization of the technological equipment in handling fissile materials one should take into consideration some additional data. The most important of these are:

- concentration of fissile material in the solution; the effect of the shape of a technological vessel (cone, cylinder, etc.) and various reflectors (water, steel, etc.) on the M_{cr} value;
- uniform distribution of plutonium and various impurities contained in the solution over the height of the vessel and the presence of a fissile material in the precipitate and on the vessel walls;
- possible surplus of the fissile material being loaded in technological vessels (by 2–3 times) over the admissible limits due to operator's faults and unforeseen circumstances (leaking valves, analytical errors, etc.);
- effects of loaded fissile materials on the critical parameters of adjacent equipment or tanks with final products.

Table 26 Critical volumes and masses of ^{235}U and ^{239}Pu water solutions in spherical vessels with uranium reflectors.

Critical parameter	^{239}Pu, g/l		^{235}U, g/l		
	38.4	316	92.2	184.3	368.7
Volume, l	13.3	11.1	11.7	9.75	9.5
Mass, g	510.0	3180.0	1080.0	1800.0	3510.0

Similar data are needed in handling fissile materials occurring as metals or other compounds.

V.S. Emelyanov [11] recalls that to implement the production of metal ^{235}U articles destined for the atomic weapons they had to produce uranium gas, uranium hexafluoride, at the diffusion plant and convert it into uranium metal. The critical mass value advised by Kurchatov to Emelyanov prior to his departure for the plant was less than the theoretical value by a factor of ten. He did this for fear that any of the managers might take his own decision to increase the ^{235}U load by a few times when checking the efficiency of the technological process: "We had to add a gaseous uranium compound to the solution and produce a solid precipitation. A nuclear fission process might start in this precipitate. This might not be an explosion-type process, ... yet, an intense neutron flux might irradiate us."

Taking rather primitive precautions in discharging gas from a cylinder to the solution, Emelyanov reduced the duration of the process by committing only a fourfold surplus of the critical mass of the loaded ^{235}U, specified by Kurchatov, and a spontaneous reaction did not occur. That was the way of leading fissile materials in the process equipment in the absence of experimental data [11].

In 1951, using the test stand in *Berezki* (Chelyabinsk-40), the first values of critical masses were experimentally determined not only for diluted homogenous assemblies but also for heterogenous systems consisting of plates made of fissile materials. The hydrogen–uranium ratio was in the range of 2.6–52. The critical ^{235}U volume ranged from 2.6 to 3.7 l and the critical mass between 9 and 1.2 kg [144].

Later, as new reactors were loaded with the uranium slugs enriched in ^{235}U, test stands were constructed at Center No. 817 under the guidance of E.D. Vorobiev[1], on which the Center workers determined critical masses for isotope reactor AI fueled with magnesian-ceramic slugs with a 2-percent ^{235}U enrichment. In 1956, based on the experiments made in Center No. 817 with fuel slugs made as plates with a 90-percent ^{235}U enrichment and various uranium to water ratios, tentative M_{cr} values were determined for reactor SM-2 being constructed at the NIIAR Institute. Until 1958 the Center proceeded with experiments aimed at determining M_{cr} for uranium solutions with a 90-percent enrichment and plutonium solutions in technological vessels of various shapes. In these experiments they studied the efficiency of heterogeneous cadmium neutron absorbers. Subsequently, experiments were conducted at other organizations and institutes using specially designed test stands in compliance with all nuclear and radiation safety regulations existing at that time.

In a handbook by Dubovskiy *et al.* [64] the results of Soviet and foreign researchers were systematized and minimum critical masses were given for ^{233}U, ^{235}U, and ^{239}Pu for water systems (Table 27). Data presented in Table 27 were obtained for a water reflector about 10 cm thick.

In nuclear reactors ^{233}U is produced from natural thorium by the reaction:

$$^{232}Th_{90} + n \rightarrow {}^{233}Th_{90} \xrightarrow{\beta^-} {}^{233}Pa_{91} \xrightarrow{\beta^-} {}^{233}U_{92}.$$

The half-lives of ^{233}Th and ^{233}Pa are 23.5 min and 27.5 days, respectively.

At a radiochemical plant they recover ^{233}U from thorium slugs, irradiated in a nuclear reactor, by a specially designed technology, much as in the case of plutonium.

When choosing the safe sizes of equipment and the safe amounts of fissile materials, one should take into account the interference of these materials enclosed in adjacent vessels or pipelines and, also, the presence of neutron absorbing impurities. The M_{cr} values presented in Table 27 were obtained for spherical vessels. In the case of a cylindrical shape, M_{cr} depends both on the concentration of fissile materials contained in solutions and on the diameter of the vessel. Each diameter is characterized by specific minimum values of the critical mass and volume depending on the concentration of uranium or plutonium in the reactor core.

Table 27 Minimum values of critical aqueous solution parameters.

Critical parameter	^{235}U	^{233}U	^{239}Pu
Minimum concentration, g/l	11.0	10.9	9.37 [64]
Critical mass, g	820	590	510
Diameter of infinite cylinder, cm	13.7	11.2	12.4
Thickness of infinite plate, cm	4.3	3.0	3.3
Critical volume, l	6.3	3.3	4.5

Table 28 Critical parameters of fissile substances.

Critical parameter	^{235}U	^{233}U	^{239}Pu	
			α-phase	δ-phase
Mass of sphere, kg	22.8	7.5	5.6	7.6
Diameter of an infinite cylinder, cm	7.8	4.8	4.3	5.3
Thickness of an infinite plate, cm	1.5	0.7	0.6	0.7

Critical parameters for the systems of the metallic compounds of fissile materials and a water reflector are given in Table 28. The uranium density is 18.8 g/cm^3, and the plutonium density is 19.6 g/cm^3 for an α-phase and 15.8 g/cm^3 for a δ-phase.

The critical volume of a sphere made of plutonium metal and equipped with a water-base neutron reflector is merely 0.3 l [64]. In the case of beryllium reflector, however, the minimum critical mass of plutonium metal is half as small (density of beryllium is equal to 1.84 g/cm^3 and that of plutonium 19.25 g/cm^3) and makes up 2.47 kg.

In a system without neutron moderator the critical mass of a fissionable isotope varies with the squared density in the case of simultaneous equal variation of the core and reflector density values and in the case of the varying density of the core in the absence of a reflector [64].

All of the above data indicate that to provide safe technological processes at NPPs, bases of the Defense Ministry, naval ships and bases, plants and institutes, which deal with fissile materials, one should know critical masses and provide adequate control of the distribution of these materials. Control should be provided throughout all phases of the technological process in each unit, settler, storage facility and tank (in the course of storage and transportation), and even in ventilation systems, and, undoubtedly, in waste solutions and warehouses half products and finished articles made of uranium and plutonium isotopes are kept.

In 1958–1959 a special laboratory was set up in the FEI Institute for elaborating standard rates for fissile materials loaded into various technological units (transportation containers, storage facilities, etc.) to ensure nuclear safety. B.G. Dubovskiy was appointed first director of this laboratory. The laboratory was equipped with necessary stands to determine the critical masses of various compounds of fissile materials. About the same time I.V. Kurchatov, Scientific Director of the atomic problem set up a Nuclear Safety Department at the Institute for Atomic Energy headed by T.N. Zubarev.

Most dangerous situations occur where nuclear reactors operating at the rated output become uncontrollable.

Nuclear and Radiation Safety during Reactor Exploitation

In all operations with fissile materials destined for nuclear weapons the amounts of these materials loaded in particular equipment used in various technological processes usually did not exceed a minimum critical mass (M_{cr}^{min}). On the contrary, in nuclear reactors (especially in high-power NPP reactors) the total mass of fissile materials loaded in the reactor core sometimes exceeded the critical mass M_{cr}^{min} by a factor of 10. So the control of a nuclear reactor has became one of the most important problems along with the problems of heat removal and the efficient utilization of nuclear fuel, i.e., an optimized fuel burnup. A specific character of a nuclear reactor exploitation is related to the presence of large amounts of radioactive substances in it. Even in low-power research reactors the amount of fission products is as high as millions of curies.

For instance, the activity of a 250-megawatt reactor is about 10^9 Ci immediately after it is shut down. Because 1 g of radium emits 1 Ci, the radioactivity of a reactor with a thermal power of several gigawatts is equivalent to that of thousand tons of radium. That is why the core destruction of any reactor always presents an extreme hazard [159].

A great body of publications is devoted to the accident at the Chernobyl NPP (ChNPP) which involved the core destruction. The most objective causes and effects of the Chernobyl accident are presented in a fundamental monograph [175] published in 1996, on the occasion of the 10th anniversary of the accident. Consider some of the less known accidents that took place in the Soviet reactors (in the early operation period) and abroad.

It is commonly supposed that the radioactivity leakage from a reactor is caused by the following three types of emergency situations:
- supercritical nuclear reactions and loss of the reactor control;
- melting of some reactor components and fuel elements caused by a delayed residual heat after the end of a chain reaction;
- possible exothermal chemical reactions involving the construction materials of a reactor.

Causes of these emergency situations were discussed in 1955 at the Geneva International Conference on the Peaceful Use of Atomic Energy. Emergency situations of all three types and various severities occurred in the Soviet production reactors, especially in those of Center No. 817, during their early operation periods. For instance, an error was made while replacing the CPS rods in the AI reactor. In the automatically controlled reactor (operating at a 1-percent power), a technician of the C&MI service manually pulled out the central rod from the core instead of an absorber rod which was to be replaced. An instantaneous uncontrollable burst of power occurred, yet, the neutron absorber rod immediately dropped into its cell, and the process was shut down. No destruction of uranium slugs or radioactivity release took place, and the personnel was not exposed.

An emergency situation occurred in another reactor, of the OK-180 type, where irradiated uranium slugs were being pulled out of the core and got jammed in a fuel channel. They were fused with material of the reactor refueling loop due to the residual heat. The reactor was shut down. Special emergency operations were developed to eliminate the consequences of this accident. Experts of Center No. 817 and other organizations took part in these operations. The experience of equipment dismantling gained in this reactor was described in [117].

Cases of intense heating (fusion) of uranium slugs in the reactor channels and excavator buckets took place in the first production reactor [160].

Emergencies involving exothermal chemical reactions in the material of the reactor core were rather common. The phenomena of uranium-graphite fusion, so-called "cakes," were encountered almost in all of the uranium-graphite reactors in their early operation period. Some examples of fusion between uranium and other core materials were reported in the publications devoted to the accidents that took place at the Beloyarsk NPP in 1964–1979.

The problem of radiation safety was given much attention since the very first years of the atomic industry. Methods and techniques used to ensure radiation safety were permanently advanced as more experience was gained. Dismantling of the core in a uranium–graphite reactor intended for tritium production was one of the most hazardous and difficult operations. This unique operation aimed at the partial replacement of a graphite moderator was implemented in the AI-reactor of Center No. 817 in the early 1950s. The AI-reactor was a graphite assembly with vertically arranged fuel channels. Fuel channels and the jackets of uranium slugs and targets used to produce isotopes were made of aluminum. Water coolant was delivered from above and, at the bottom, in the fuel channel outlet, it had a temperature of 80–90°. The rated capacity of the reactor was 50 MW. As noted in [161], the reactor was designed for the production of an important product, tritium. The reactor fuel was based on an enriched uranium metal. Uranium metal containing grains of 2% ^{235}U was uniformly distributed over a metal matrix made of magnesium. Being a poor neutron absorber, magnesium was chosen by LIPAN physicists who designed and produced specifications for the reactor core. A technology for manufacturing dispersion uranium–magnesium slugs was devised by R.S. Ambartsumyan and A.M. Glukhov. Slugs of magnesium ceramics were subject to a preliminary testing in reactor A of Center No. 817. However, as the slugs were loaded in the AI reactor and the reactor was put into operation (December 1951), they started to disintegrate because of the failure of their aluminum jackets. Theoretically the dispersion slugs were ideal because magnesium did not absorb neutrons and the "state" neutrons were used in the reactor with a high efficiency. The expression: "Save the state neutrons!" was often used by I.V. Kurchatov and V.S. Fursov, his deputy. Out of the 248 cells of the reactor only 140 were charged with these uranium slugs. The total weight of uranium in the AI reactor was 3 tons. Soon after the start-up, water penetrated into the cores of the fuel elements through the collapsed jackets of the uranium slugs. This resulted in the destruction of the slugs, burnup of magnesium, and the fusion of magnesium and uranium with the graphite. This in turn resulted in the radiation contamination of the reactor premises and the basic technological water basin, Lake Kyzyl-Tyash. Because the reactor was charged and discharged via its top, even single cases involving the failure of the uranium slugs resulted in the contamination of the central hall and other premises of the reactor. A specific feature of this research and production reactor was that its design assumed the possibility of its dismantling. Though the reactor operated at a medium output (50 MW), the maximum density of thermal neutron flux was as high as 4.5×10^{13} cm$^{-2} \cdot$ s^{-1}. The temperature of graphite in the middle of the reactor core was as high as 400–500°. This resulted in the burnup of graphite, especially when air penetrated into the graphite stack.

Radiation Safety during the AI Reactor Rehabilitation

The collapses of individual uranium slugs caused material defects in the graphite stack of the reactor, which complicated the radiation situation. The authorities of the Medium Machine Building Ministry supported Kurchatov's proposal to dismantle part of the reactor, examine the characteristics of the altered graphite, and replace the collapsed bricks ($200 \times 200 \times 600$ mm) of the graphite stack. The basic job was done by the servicing personnel of the reactor in 1956. Measurements of alterations and collapses in the graphite bricks were made with the assistance of workers from the Central Laboratory and Laboratory No. 5 of the Center. A major contribution to examining the properties of the irradiated graphite was made by V.I. Klimenkov, head of the laboratory team. Subsequent to the reactor discharging and the removal of graphite columns, gamma radiation was measured in particular compartments of the reactor site under the guidance of B.M. Dolishnuk, director of the dosemeter services. After the discharging of uranium the exposure rate at the reactor site was as high as 200 R/h [146].

The managers of Plant No. 156 (which included the AI reactor), Director F.Ya. Ovchinnikov and Chief Engineer B.V. Brokhovich, had good experience in implementing operations under conditions of intense radiation. Nearly all of the operations dealing with the partial replacement of the graphite stack at the AI reactor were conducted remotely as it had been done during similar operations at the first production reactor of Center No. 817 and other reactors. All of them, particularly Ovchinnikov, Brokhovich, and the personnel of duty shifts and AI reactor services, provided well coordinated operations under intense radiation conditions. This reactor dismantling was the first experience of this kind in the world. Despite the high radiation, the average exposure of the personnel engaged in these operations was within an admissible limit which at that time was 0.05 R per working day.

Overall 40 graphite columns containing approximately 400 bricks in total were removed. Bricks taken from 21 columns which were located in the affected domains of the graphite stack were examined in detail. Besides alterations in the shape of the bricks, they examined changes in the heat conductivity of graphite, its electric resistance, and strength as a function of the neutron flux density, neutron energy, and graphite temperature. Upon the completion of the reassembly operations and the examination of graphite stack, the AI reactor successfully operated until 1987. It was completely decommissioned on May 25, 1987, i.e., 36 years after its start-up.

First Governmental Radiation and Nuclear Safety Agencies

A State Radiation Safety Monitoring Service was set up in the USSR prior to the start-up of experimental reactor F-1 at Laboratory No. 2. Leading institutes of the Academy of Sciences and the Ministry of Health were engaged for devising the radiation monitoring methods and instruments and drawing norms and rules to regulate the handling of radioactive materials. Based on the experience gained in handling radium and at cyclotrons, they devised admissible personnel exposure rates and admissible rates for disposing of radioactive elements into the hydrosphere and atmosphere. The first session of the Scientific and Technical Section of the PGU held on April 24, 1946, approved a proposal made by Ya.B. Zeldovich to set up an individual photographic monitoring of irradiation hazard. Zeldovich was charged with setting up this monitoring at Laboratory No. 2. Concurrently with preparation for the start-up of experimental reactor F-1 at Laboratory No. 2, specialists from the Ministry of Health and the PGU set up a radiation monitoring system at the first Soviet enterprises dealing with radiation sources and natural uranium. It was decided to start manufacturing of dosimeters in the PGU system. Based on the reports by A.A. Letavet and L.A. Orbelli, it was decided that (1) medical and sanitary servicing of the RIAN institute and Laboratory No. 2 should be provided by the Institute of Occupational Diseases, Ministry of Health (Director A.A. Letavet), and (2) medical units should be arranged at the PGU enterprises.

On September 18, 1946, the Section adopted a resolution based on the results of the medical examination of workers from Center No. 6 mining uranium ore in Central Asia.

Kurchatov and Panasyuk [26] mentioned that a radiation monitoring group headed by B.G. Dubovskiy had been set up long before they started building a nuclear pile at Laboratory No. 2. Apart from the development of gamma-ray, neutron, and radioactive gas dosimeters, this group maintained relations with biologists and medical men who were examining the effect of radiation on man. "By the time we started up the pile we already had experimental units of a remote gamma-ray dosimeter to detect radioactive gas in the air. Moreover, researchers from the Radiation Laboratory of the USSR Academy of Medical Sciences devised an individual integral dosimeter using thimble ionization chambers and photographic film. With the aid of these devices we were monitoring radiation and conducted biological experiments with animals. No severe injuries were found among the personnel servicing the first Soviet atomic pile. Somewhat later, dosimeters were devised for detecting thermal and fast neutrons [26]."

Some other institutes and enterprises were involved in designing dosimeters. Having heard reports on designing dosimeters at the RIAN and LFEI, Section 5 decided to include their projects into a consolidated program on the dose monitoring problem, and charge A.I. Burnazyan (Deputy Minister of Health) with controlling the implementation of this program. In the minutes Nos 16-17 of a section meeting that took place in December 1946, the section approved the proposals of B.G. Dubovskiy and I.S. Panasyuk, researchers from Laboratory No. 2, German scientists who worked at Laboratory B, South Urals (group headed by K. Zimmer), and at Plant No. 12, Elektrostal, were also recruited for manufacturing dosimeters.

Radiologists were trained to work at the PGU enterprises. Dose monitoring services were set up at all nuclear facilities. Later, the functions of studying radiation effects on the human organism and, also, developing norms and rules regulating manipulations with radioactive materials were assigned to the Institute of Biophysics.

The State Radiation Safety Service headed by the USSR Ministry of Health worked out Radiation Safety Standards (NRB) and Basic Sanitary Regulations (OSP). These standards were continuously revised and improved. In the third edition of NRB-76/87 and OSP-72/87 it was pointed out that these standards covered all of the enterprises and institutions of the Ministries and Departments dealing with the production, processing, utilization, storage, reprocessing, disposal, and transportation of natural and artificial radioactive materials and other sources of radiation [81]. These Standards stated that any departmental or sectoral regulations and manuals should not conflict with the state standards. Those found to be guilty of violating the standards should be brought to an administrative or even a criminal trial.

There was no officially approved state service responsible for nuclear safety in the USSR until 1972. However, in 1958 a departmental research laboratory of the Minsredmash Ministry was set up at the FEI institute. It was equipped with necessary test stands to conduct experiments on determining conditions for ensuring nuclear safety in the atomic industry. The laboratory was headed by B.G. Dubovskiy, one of Kurchatov's workers, a participant of the F-1 reactor start-up, the former scientific director of the production reactor A. Besides experimental determination of critical masses, his laboratory was drawing main standards regulating nuclear safety at the enterprises of the atomic industry. At all of these enterprises they set up a position of an authorized officer (Deputy Chief Engineer or Chief Physicist) who was to control the adherence to the above-mentioned standards. In fact, Dubovskiy was a scientific supervisor of these problems at his Laboratory and in the atomic industry as a whole.

The development of nuclear power industry, commissioning of research reactors, and construction of nuclear ice-breakers posed nuclear safety problems in various branches and departments of the national economy. After a number of accidents involving the spontaneous nuclear chain reactions took place at some production enterprises and research institutes, the first State Atomic Inspection Agency (*Gosatomnadzor*) was set up in the Minsredmash Ministry, the basic manufacturer of fissile materials. This agency was charged with nuclear safety supervision. N.I. Kozlov,[2] one of the leading experts in the operation of nuclear reactors, was appointed Director of the Gosatomnadzor Agency.

Though this inspection agency was charged with the functions of state nuclear safety control, because of its subordination to the administration of the Minsredmash Ministry, its activities could not be independent to a full extent. Any radical resolutions issued by N.I. Kozlov on particular violations of nuclear safety and measures to be taken to eliminate them in various research institutes, design bureaus, and construction enterprises of the Minsredmash Ministry could be blocked by the Minister or his First Deputy. Because of this, it was decided to make this agency independent. It was declared that in addition to nuclear safety, this agency should also control particular problems of engineering safety concerning the operation of major equipment used in NPP nuclear reactors.

To provide the state control of nuclear and engineering safety, a decree No. 653-27 issued by the Council of Ministers of the USSR on July 14, 1983, set up the State Atomic Energy Inspection Agency of the USSR.[3] Its basic functions were:

- supervise all ministries, departments, enterprises, organizations, institutions, and officials in their keeping to the existing regulations, standards, and instructions on nuclear and engineering safety when they design or exploit facilities of the nuclear power industry;
- supervise the designing and manufacture of equipment destined for these facilities, and also the storage and transportation of nuclear fuel and

radioactive wastes at them;
- supervise the nuclear safety of operations conducted at experimental and research reactors at research institutes, design bureaus, and navy and merchant power units;
- control activities in drawing standards on the safe operation of nuclear power facilities carried out in various ministries and departments;
- control the registration of fissile materials at nuclear power facilities;
- control measures undertaken to prevent accidents at nuclear power facilities and the abilities of enterprises to eliminate them;
- supervise and control the quality of building operations and equipment installation at nuclear power facilities.

Among other tasks assigned to the State Atomic Energy Inspection Agency were the coordination and supervision of scientific researches conducted by different departments to substantiate requirements to the safety of nuclear power facilities and the efficiency of engineering measures taken to ensure safety at them. Regulations [169] stated that the Kurchatov Institute for Atomic Energy was charged with the scientific supervision of research into the safety of nuclear power facilities. The USSR Ministry of Health was charged with the supervision of radiation safety problems, and the Defense Ministry with the supervision of the registration and storage of fissile materials used in nuclear weapons production.

After the USSR breakdown, each independent state had instituted its own safety inspection body. For instance, in Russia these functions were assigned to the State Atomic Inspection Agency (Gosatomnadzor). However, the interdepartmental distribution of responsibilities for particular functions of the state inspection of engineering and radiation safety at the enterprises belonging to the Minatom Ministry of Russia and the radiation safety and storage of radioactive materials at various facilities and bases of the Defense Ministry is still being specified. The dismantling of nuclear armaments and the reduction of nuclear weapons make these problems more urgent both for Russia's enterprises and those of other CIS states possessing articles of enriched ^{235}U and plutonium.

At the Board meeting of the Minatom Ministry (Ministry of Atomic Industry) which took place on October 5, 1993, they discussed the modern state of nuclear and radiation safety at the enterprises of the industry and concluded that the following nuclear- and radiation-hazardous facilities were under operation at that time:
- 29 power units located at 9 NPPs with a total rated yield of 21,200 MW;
- 15 research reactors and 30 critical test stands located at research institutes and design bureaus;
- 3 production uranium-graphite reactors of the ADE, *Ruslan*, and *Ludmila* types;
- plants producing and processing nuclear fuel (2 plants producing fuel elements and assemblies, 4 isotope separation plants, 2 chemico-metallurgical plants, and 3 radiochemical plants);

- nuclear-hazardous sites and laboratories at the research institutes;
- facilities and sites destined for storing enriched and spent nuclear fuel and fissile materials at NPPs, production and research reactors, fuel-producing plants, research institutes, and design bureaus.

The 1993 radiation situation at the enterprises of the atomic industry was as can be characterized by the following data. The average external radiation dose received by personnel was 0.47 rem as compared to the admissible radiation dose of 5 rem/year. 25 workers of the Smolensk NPP received up to 7.5 rem when doing some repair jobs (within the limits permitted by the bodies of the State Sanitary Inspectorate). 49 miners of the Argun Mining and Chemical Center were found to receive an insignificant surplus over the rated yearly intake of radionuclides.

Because of a possible switch to new, lower radiation dose limits recommended by the International Atomic Energy Agency and International Commission for Radiological Protection of personnel and population (2 rem/year and 0.1 rem/year, respectively), they planned to take a large number of scientific, engineering, and organizational measures to provide implementation of the new dose rates.

The Ministry Board noted that dual-purpose uranium-graphite reactors ADE-2 (Mining and Chemical Center, Krasnoyarsk-26), ADE-4 and ADE-5 (Siberian Chemical Center, Tomsk) were operating steadily and supplied sufficient heat and electric power for the secret towns and that Siberian NPP provided 75% of the heat supply for the Tomsk regional center.

Following a governmental decree, a resolution was adopted by the Ministry Board to draw up a Federal Program "Safety of the Atomic Power Industry of the Russian Federation." In addition, other decisions were made to ensure nuclear and radiation safety, including the development of principles for the mitigation of emergency situations in the atomic industry.

Apart from the Ministry of Atomic Industry which deals with its own problems, national agencies play an important role in providing nuclear and radiation safety in Russia. The functions of these agencies are to be agreed upon, and the Law on the Atomic Energy is to be approved in the near future.

NOTES

1   In the latest years he worked for the Joint Institute for Nuclear Research, Dubna.
2   N.I. Kozlov began his career in 1948 at the first production nuclear reactor. Later, he worked as Director of the uranium-graphite reactor AV-2 and Deputy Chief Engineer of Center No. 817 in Chelyabinsk-40.
3   E.V. Kulov was appointed as Director of *Gosatomenergonadzor* Agency (independent of the Minsredmash Ministry) and V.A. Sidorenko as his first deputy.

Conclusion

The utilization of nuclear energy became possible firstly due to the discovery of radioactivity; the centenary of this breakthrough was celebrated by the international scientific community in 1996. Foundations for radioactivity studies and nuclear fission physics in the USSR were laid in the pre-war period by V.G. Khlopin, L.S. Kolovrat-Chervinskiy, G.A. Gamow, L.V. Mysovskiy, A.F. Ioffe, I.V. Kurchatov and many other scientists. Great scientific advances such as radium production and its use in medicine, nuclear physics, and radiochemistry, the construction of a first cyclotron at RIAN (first in Europe), and the discovery of uranium fission enabled Academician V.I. Vernadskiy to establish in 1940 a Uranium Commission at the USSR Academy of Sciences. In 1940–1941 this Commission proposed the first version of the State Program on the use of uranium fission energy. This program stipulated activities in the exploration and mining of uranium resources, appointed the leading research institutes responsible for the development of uranium isotope separation techniques, and proposed that some of the institutes should be provided with new experimental equipment to carry out research into the utilization of nuclear fission in the national economy. Just before the war, two young researchers, G.N. Flerov and K.A. Petrzhak, discovered a spontaneous (without neutron irradiation) fission of uranium. Two monographs published by the Leningrad physicists, one by theoretician G.A. Gamow, the other by experimenter I.V. Kurchatov, demonstrated the abilities of Soviet scientists to carry out fundamental research into nuclear fission. In 1939–1940, Ya.B. Zeldovich and Yu.B. Khariton calculated that a nuclear chain reaction could be obtained in a system consisting of uranium, slightly enriched with ^{235}U, and heavy hydrogen or graphite. They also specified conditions for a nuclear explosion and evaluated its demolition power. Information about the accumulation of considerable reserves of uranium in the USA and Germany and publications on the determination of the yield of secondary neutrons per primary neutron captured by a uranium nucleus (E. Fermi and F. Joliot-Curie in the West, G.N. Flerov and L.I. Rusinov in the Soviet Union) proved the reality of a nuclear chain reaction. After the beginning of the war, publications of this kind were banned in the West; in the USSR many of the researchers engaged in this field joined the Army; the leading research centers of Moscow and Leningrad were moved to the East, and research into the utilization of nuclear fission was suspended and renewed after getting intelligence data evidencing that Germany, UK and the United States were very active in the use of nuclear energy for military purposes. Activities of the Soviet intelligence which informed the government of secret nuclear research under way abroad was well organized. As early as mid-1943, scientific coordinator of the Uranium Project I.V. Kurchatov, in his report to M.G. Pervukhin, emphasized the great value of 237 documents on ^{239}Pu and enriched ^{235}U production provided by the intelligence. This is how Academician Yu.B. Khariton evaluated intelligence information in his book [170]: "Intelligence helped our researchers to save time and avoid "fall flats" when conducting the first A-bomb test which was of great political significance. Our intelligence helped to make I.V. Kurchatov one of the most informed nuclear physicists in the world. Aware of the advances of his colleagues in the USSR, he at the same time was aware of the progress of Western scientists. This was of great importance in the initial stage of the nuclear weapons race." It should be noted that Kurchatov (given Pervukhin's authorization) had the right of showing these documents to his colleagues who were his deputies in particular areas of the Uranium Project. Intelligence data were of great value not only to nuclear physicists. They were equally important to scholars and designers of equipment and technologies for uranium isotope separation, uranium/graphite and heavy-water nuclear reactors fuelled with natural uranium, for the development of uranium metal, uranium hexafluoride, heavy-water production technologies, etc. Moreover, the intelligence provided documents on the bomb design. Naturally, Kurchatov understood that to evaluate these documents, he needed not only nuclear physicists but also experts in metallurgy and metal physics who worked with metal plutonium and uranium, experts in physical chemistry, and other researchers who headed particular areas of the Soviet atomic project. Therefore, not only Kurchatov himself, but I.K. Kikoin, Yu.B. Khariton A.A. Bochvar, L.A. Artsimovich and other leading scientists were well informed on the advances of their Western counterparts [102, 171]. Nevertheless, as mentioned by the US researchers, scientific aspects were not key problems of any atomic project. In 1945, G. Smyth published his monograph *Atomic Energy for Military Purposes* (an official report of the US Government on the atomic bomb project) in the United States. This publication indicates that the US authorities were sure that because of great technical difficulties Soviet industry would not be able to start production of plutonium and enriched ^{235}U and build appropriate production facilities, and hence the USSR would not be able to develop an atomic bomb in less than 10 years [36]:

"Basic difficulties the Soviets have to overcome will be related to the current state of their heavy industry and production. The Soviet Union has brilliant scientists who are able to find answers to all the questions by themselves. Moreover, they got much help from the intelligence."

The mobilization of the industrial and human resources of the country and the cooperation of many branches of industry in the work on the Uranium Project totally disproved the US attitude toward the abilities of the Soviet industry. The Uranium Project proved a success not only due to the top priority given to the program and the leadership of prominent scientists under the guidance of Kurchatov, but also due to the attention the authorities of the country gave to the project. Starting from 1945, four deputies of the Chairman of the People's Commissars Council of the USSR (L.P. Beriya, M.G. Pervukhin, V.A. Malyshev, and N.A. Voznesenskiy) were personally responsible to Stalin for its success. They were responsible for the timely allocation of resources to the industry, R&D centers, research institutes, and production facilities of the PGU, which later formed the backbone of the newly-born Soviet atomic industry.

Though domestic researchers and engineers were able to use intelligence information, the basic techniques and technologies were developed originally. Many research institutes of the USSR Academy of Sciences were involved in top-priority scientific activities to verify the intelligence data, find the missing ones, and develop new technological processes, control systems, and monitoring equipment needed for fission materials production and the development of a bomb design.

The date of September 30, 1995, was the 50th anniversary of the Soviet atomic industry. The most important contributors to the creation of the nuclear shield for the country were those who were responsible for the first facilities where fissile materials were produced. Though these facilities were few in number, many research centers, design bureaus, engineering organizations, building contractors, and subcontractors from different industrial branches were involved in the project. Tens of thousands of engineers and technicians of various professions participated in their building. All activities were coordinated by the Special Committee and PGU administrators and personally by Kurchatov who headed the Uranium Project in the post-war period.

A group of people engaged in the plutonium bomb development, who were decorated in 1949, included the Head of Laboratory No. 2, the KB-11 designers, the Chief Engineer of Center No. 817, the heads of some research programs, and the Special Committee and PGU administrators. The title of Hero of Socialist Labor was conferred on more than 20 persons. This group included:

- M.G. Pervukhin − Deputy Chairman of the USSR Council of Ministers;
- I.V. Kurchatov − Scientific Supervisor of the Project and Center No. 817;
- E.P. Slavskiy − Chief Engineer of Center No. 817;

- V.G. Khlopin — Director of the RIAN Institute;
- A.A. Bochvar — Head of NII-9 Department, Scientific Supervisor of Plant V, Center No. 817;
- A.P. Zavenyagin — First Deputy Director, PGU;
- Yu.B. Khariton - Scientific Supervisor, KB-11;
- K.I. Shchelkin — Deputy Scientific Coordinator, KB-11;
- A.N. Komarovskiy - Deputy Director, PGU; Director of the Glavprom-stroy Directorate, NKVD;
- P.K. Georgievskiy - Deputy Director, Glavpromstroy Directorate, NKVD.

B.L. Vannikov and B.G. Muzrukov received their second Hero of Socialist Labor medals. By a decree of October 29, 1949, several other top managers of the industry and researchers participating in the production of fissile materials and in the designing and building of the plutonium bomb became Heroes of Socialist Labor. Among them were V.I. Alferov, G.N. Flerov, N.L. Dukhov, A.P. Vinogradov, Ya.B. Zeldovich, A.N. Kallistov, N.A. Dollezhal, B.N. Chirkov, Yu.N. Golovanov, German physicist N. Riel (who worked at Plant No. 12) and several others. Later the Soviet Government decorated employees of various industrial enterprises, research centers and design bureaus, engaged in the design and production of materials and components, equipment and instruments used for the nuclear weapons development and for utilization of nuclear technologies in the national economy. Stalin Prizes[1] were awarded to a large group of employees from numerous research institutes, design bureaus and production facilities, who were involved in Program No. 1, a project for manufacturing the first Soviet plutonium bomb.

In 1951, by a decree No. 4964-2148 of December 6, the title of Hero of Socialist Labor was awarded for setting up the ^{235}U production (with a 90-percent enrichment) and manufacturing nuclear weapons from this uranium, to the following persons:

- I.K. Kikoin — Scientific Supervisor of Center No. 813, designer of a gas diffusion uranium enrichment technique;
- S.L. Sobolev — Deputy Scientific Supervisor of the problem;
- A.I. Churin — Director of Center No. 813.

I.V. Kurchatov, Yu.B. Khariton, and K.I. Shchelkin were decorated with their second gold Hammer and Sickle medals. Also all of them, and M.P. Radionov, Chief Engineer of Center No. 813, were awarded first-grade Stalin Prizes.

All in all, Stalin Prizes of different grades were awarded to 390 people from 55 organizations.

Many employees of Center No. 813, researchers from Laboratory No. 2 of the Academy, and their workers from various industrial branches were decorated with orders and medals.

The international situation of that time called for enormous efforts from the national economy to provide an extensive development of the atomic industry. The administrative and management structure of the developing atomic industry was subject to perpetual improvement in 1949-1953. B.L. Vannikov was moved to the Special Committee as Beriya's First Deputy, and A.P. Zavenyagin was appointed Head of the PGU. E.P. Slavskiy left Center No. 817 and returned to the PGU as First Deputy Director. The Second and Third Directorates were set up with the USSR Council of Ministers, headed by P.Ya. Antropov and V.S. Ryabikov, respectively.

On June 26, 1953, a Ministry of Medium Machine Building (Minsredmash) was established, which united the First, Second, and Third Directorates of the USSR Council of Ministers. A special building was erected in Moscow in the Bolshaya Ordynka street to accommodate these Directorates. At present the Russian Ministry of Atomic Energy (Minatom) resides in the same premises.

V.A. Malyshev, a well known statesman and mechanical engineer, was appointed the first Minister of Medium Machine Building. Concurrently, he held a position of a Deputy Chairman in the USSR Council of Ministers. Since late 1945 this man was a pioneer of many research projects and coordinated the efforts of all machine-building enterprises engaged in manufacturing equipment for the gas diffusion plants producing enriched ^{235}U. In 1953 B.L. Vannikov and V.M. Khrunichev were appointed Deputy Ministers. Along with I.V. Kurchatov, A.P. Aleksandrov, A.I. Leipunskiy and several others, V.A. Malyshev was responsible for the installation of the first nuclear reactors in the Soviet Navy and merchant vessels (icebreakers).

In 1953 the Minsredmash Ministry took most of the responsibilities of the Special Committee. By a Decree of the Council of Ministers issued on July 9, 1953, basic divisions of Glavpromstroy (which was at that time under the USSR Ministry of Internal Affairs) were incorporated into the Minsredmash structure.

Basic developments took place in the industry under E.P. Slavskiy, who headed Minsredmash for almost 30 years starting from 1957. In these years new R&D organizations were created, new funds and personnel were allocated to the construction enterprises of the ministry, a number of new manufacturing plants and production facilities were built, which promoted the development of numerous defense and civil nuclear projects.

Gold, zirconium, and fertilizer production plants were among the largest in the industry. The construction of bulk chemicals production facilities, especially sulfuric acid and special chemicals production plants, as well as plants for building production, enabled the enterprises of the Ministry to become self-sufficient with a minimum of resources derived from other sectors of the country's economy. As stated by V.N. Mikhailov [172] who at present heads the Ministry: "Minatom produces the purest gold in the world. We have the most cost-effective zirconium-production technologies, the most effective technique for isotope separation which consumes 20 times less energy than its

American rivals. ...We produce pure molybdenum, tungsten, and vanadium oxides from low-grade uranium ore. Minatom has several factories for producing sulfuric, nitric and hydrofluoric acids and elementary fluorine. Our enterprises produce niobium and tantalum, zirconium and hafnium, lithium and beryllium, alkaline-earth metals and manufacture articles from these metals." However, throughout the previous years, the defense order for manufacturing nuclear and thermonuclear weapons was the most important job performed by the atomic industry. At present, the industry faces an extremely difficult problem of dismantling and annihilating a considerable portion of the existing nuclear weapons stock. Difficulties arising in the solution of this problem, the experience possessed by Russia's Minatom Ministry and departments, and necessary investments are described by V.N. Mikhailov [172] who states: "A nuclear weapon is a sophisticated device which incorporates electronics, initiators, and fissile materials − uranium, plutonium, and tritium, as well as conventional chemical explosives. Its lifetime is limited (10–15 years) and then the weapon should be dismantled. Therefore, dismantling of nuclear weapons is a kind of conventional procedure for the industry. ...As to the tactical nuclear weapons, the volume of work needed to utilize or destroy nuclear warheads will increase. Even now our facilities are more engaged in dismantling aged warheads than in producing new ones."

Not only advances and achievements marked the long way of the Soviet atomic industry. Along with accidents and nuclear disasters known internationally, first Soviet nuclear plants and facilities suffered numerous minor failures due to the imperfections of the technological processes devised by scientists and designers and to the errors of the users.

The development of the atomic industry demanded arranging medical service for the personnel. Joint efforts of the Minsredmash, the Trade Unions Central Committee, and the Ministry of Public Health resulted in the organization of medical units at most of nuclear fuel-producing enterprises; these were staffed with specially trained medical personnel and appropriate medical equipment. Hospitals and medical centers for the employees and their families were built. At present over 100 nuclear fuel production facilities (including 9 nuclear power plants, 8 nuclear fuel producing centers, nuclear weapon production plants, etc., are located in 24 administrative districts of Russia. In 1951–1980 thanks to the efforts of E.P. Slavskiy, A.I. Burnazyan, and A.N. Kallistov, sanatoria and resort centers for personnel rehabilitation were opened practically in every summer recreation area of the country.

At present, the total number of people using medical institutions of Russia's Minatom exceeds 2.2 million, including 500 thousand children. The Minatom's Board and the top officials of the Ministry of Public Health are trying to find ways to preserve the integrated system of a special medical and sanitary aid for the personnel of the atomic industry.

The fundamental research centers and excellent experimental facilities, specially designed for solving problems in nuclear and solid-state physics enabled the researchers of Minsredmash to carry out in 1970–1980 unique scientific researches on the international level. This is attested by the fact that 15% of patents annually granted in this country came from the atomic industry.

In conclusion, I would like to express much thanks to the veterans of the atomic industry A.G. Meshkov, A.M. Petrosyants, N.I. Kozlov, N.B. Karpov, V.V. Goncharov, M.G. Meshcheryakov, L.I. Nadporozhskiy, F.Ya. Ovchinnikov, B.V. Brokhovich, N.I. Chesnokov, V.I. Merkin, Ya.P. Dokuchayev, and V.G. Terentiev for helpful discussions and valuable comments.

NOTES

1 Established in 1939, these prizes were called Stalin Prizes until 1955 when they were renamed State Prizes. Lenin Prizes were introduced in 1957.

References

1. O.A. Staroselskaya-Nikitina, A History of Radioactivity and the Origin of Nuclear Physics (Moscow: Izd-vo AN SSSR, 1963).
2. F. Soddy, A History of Atomic Energy (in Russian) (Moscow: Atomizdat, 1979).
3. Khlopin Radium Institute: 50th Anniversary of Foundation (Leningrad: Nauka, 1972).
4. V.V. Igonin, Atom in the USSR (Saratov: Saratov University Press, 1975).
5. Yu.A. Khramov, Physicists: Biographical Reference, Ed. A.I. Akhiezer (Moscow: Nauka, 1983).
6. In August 1949, Rodina, Nos. 8, 9 (1992).
7. M.G. Pervukhin, How the Atomic Problem was Solved in the USSR, Ibid.
8. E.P. Slavskiy, The last interview (edited by A. Artizov and R. Usikov), Ibid., No.9.
9. A.P. Aleksandrov, Nuclear physics and development of atomic engineering in the USSR, in: October Revolution and Science Advances (Moscow: Izd-vo AN SSSR, 1967).
10. 50th Anniversary of Kharkov Physical Engineering Institute (Kiev: Naukova Dumka, 1978).
11. Memories of Igor Kurchatov (Moscow: Nauka, 1988).
12. 50th Anniversary of Modern Nuclear Physics: Collection of papers (Moscow: Energoatomizdat, 1982).
13. I.V. Kurchatov, Nuclear Fission (Moscow-Leningrad: Red. Obshetekhn. Distsiplin, 1935).
14. I.N. Golovin, I.V. Kurchatov, 2nd Ed. (Moscow: Atomizdat, 1972).
15. 80th Anniversary of Yu.B. Khariton's birthday, in: Problems of experimental and theoretical physics (Leningrad, Nauka, 1984).
16. A.L. Yanshin, Exploration and development of mineral resources in the early years of the Soviet power, in: Great October − 70th anniversary. Scientific-engineering and social advances (Moscow: Nauka, 1987).
17. L.V. Komlev, G.S. Sinitsyna, and M.P. Kovalskaya, Uranium problem, in: Academician Khlopin: Sketches and memoirs of contemporaries (Leningrad: Nauka, 1987).

18. I.V. Kurchatov, Fission of heavy nuclei, Izv. Akad. Nauk SSSR. Ser. Fiz., Nos. 4, 5 (1941).

19. A.I. Yoirysh and I.D. Morokhov, Hiroshima (Moscow: Atomizdat, 1979).

20. V. Ovchinnikov, Hot ashes. Chronicle of the atomic weapons race (Moscow: APN, 1984).

21. G.N. Flerov, Early years of the nuclear energy utilization, Moscow news, No. 16 (1985).

22. One hundred forty interviews with Molotov: Notes from F. Chuev's diary (Moscow: Terra-Terra, 1991).

23. G. Seaborg, Artificial Transuranic Elements (in Russian) (Moscow: Atomizdat, 1965).

24. V.V. Goncharov, First (primary) phases in solving the atomic problem in the USSR (Moscow: Kurchatov Institute IAE, 1990).

25. A.P. Grinberg and V.Ya. Frenkel, Igor Kurchatov at the Physical Engineering Institute (Leningrad: Nauka, 1984).

26. I.V. Kurchatov and I.S. Panasyuk, Construction and commissioning of the first Soviet uranium-graphite pile with a self-sustaining chain reaction. January–December 1946, Report No. 3498-Ts, 1947.

27. I.F. Zhezherun, Construction and start-up of the first atomic reactor in the Soviet Union (Moscow: Atomizdat, 1978).

28. N.M. Sinev, Production of enriched uranium for nuclear weapons and power industry (Moscow: TsNIIatominform, 1991).

29. Yu.B. Khariton, Were the USSR nuclear weapons brought from America or produced independently? Izvestiya, December 8 (1992).

30. Science – Engineering – Management. Proceedings of All-American Conference on the Problems of Integrated Programs in the Epoch of Scientific and Engineering Progress (in Russian) (Moscow: Sov. Radio, 1966).

31. A.D. Sakharov, Memoirs, Znamya, Nos. 10–12 (1990).

32. World War II - reminiscences by Winston Churchill, Charles de Gaulle, Cordell Hull, William Legi, and Dwight Eisenhower (in Russian) (Moscow: Izd-vo Polit. Liter., 1990).

33. Yu.N. Elfimov, Marshall of the atomic industry: Biographical sketch devoted to A.P. Zavenyagin (Chelyabinsk: Yuzhno-Ural. Izd-vo, 1991).

34. V.S. Gubarev, Arzamas-16 (Moscow: IzdAt, 1992).

35. A.N. Komarovsky, Notes of a construction-builder (Moscow: Voenizdat, 1973).

36. H.D. Smyth, Atomic Energy for Military Purposes (Princeton University Press, 1945) (Transl. from English).

37. V.S. Fursov, Research of the Academy of Sciences on uranium-graphite reactors, in: Papers presented at the Academy session devoted to the peaceful uses of atomic energy, July 1–5, 1955 (Moscow: Izd-vo AN SSSR, 1955).

38. Pages of the VNIINM history: Memoirs (Moscow: TsNIIatominform, 1993).

39. N.A. Dollezhal, Origin of the man-made world, in: Tribuna akademika (Moscow: Znanie, 1989).

40. E.P. Slavskiy, The period when this country was supported by the shoulders of nuclear giants, Voenno-istoricheskiy Zhurnal, No. 9 (1993).

41. B.V. Brokhovich, Kurchatov in Chelyabinsk-40, in: Memoirs of veterans (Chelyabinsk-40: "Mayak", 1993).

42. V.I. Shevchenko, About myself and the first reactor plant (1947–1957) (Ozersk: unpublished manuscript, 1992).

43. M.V. Gladyshev, Plutonium for the atomic bomb: reminiscences of the Plutonium plant director (Chelyabinsk-40, "Mayak", 1992).

44. Jubilee session of the Scientifiic Council of Russia's Research Center "Kurchatov Institute", Proceedings (Moscow: RNTs "Kurchatov Institute", 1993).

45. A.D. Galanin, Theory of thermal neutron reactors (Moscow: Izd-vo Glav. Upr. Isp. Atom. Ener., 1959).

46. B.V. Nikipelov, A.F. Lyzlov, and N.A. Koshurnikova, Experience of the fisrt atomic enterprise: Personnel exposure and health, Priroda, No. 2 (1990).

47. B.V. Nikipelov, A.F. Lyzlov, and N.A. Koshurnikova, Exposure rates and longrun effects, NIMB, No. 4 (1992).

48. V.N. Doshenko, All truth about radiation: Notes of a physician taken from semi-open deed boxes (Ozersk, 1991).

49. V.M. Vdovenko, Z.V. Ershova, B.S. Kolychev, and V.V. Fomin, Radiochemistry, in: Soviet atomic science and engineering (Moscow: Atomizdat, 1967).

50. Z.V. Ershova, Meetings with Academician V.G. Khlopin, in: [17].

51. M.G. Meshcheryakov, V.G. Khlopin: Ascent to the last summit, Priroda No. 3 (1993).

52. Scientific and engineering fundamentals of the nuclear power industry, Ed. K. Goodman (in Russian) (Moscow: Izd-vo Inostr. Liter., 1948, V.1; 1950, V.2).

53. V.I. Spitsyn, V.A. Balukova, et al., Migration of radioelements in soil, in: Proc. Second International Conference on the Use of Atomic Energy, Geneva, 1958 (Moscow: Izd-vo Glav. Upr. Atom. Ener., 1959, Vol. 4).

54. Ya.P. Dokuchaev, Reminiscences of testing the first Soviet plutonium bomb at 6.30 a.m. on August 29, 1949, at Semipalatinsk test site (Yaroslavl: Yaroslavl State University, 1992).

55. Views of Soviet scientists on the tests of nuclear weapons (Moscow: Atomizdat, 1959).

56. Comments on the draft law of Russian Federation "Social protection of Russian Federation citizens − victims of radiation polutions", Atom-Pressa No. 22 (1992).

57. A.V. Akleev et al., Radioactive environmental pollution in the South Ural region and its effect on the population health, Ed. Academician L.D. Buldakov (Moscow: TsNIIatominform, 1991).

58. A.K. Kruglov, Yu.V. Reshetko, and Yu.V. Smirnov, Rehabilitation of contaminated areas, Inform. Bull. No. 10 (1992) (Public Information Center for Atomic Energy).

59. V.S. Emeliyanov, That's what he was..., in: [11].

60. A.S. Zaimovskiy, V.V. Kalashnikov, and I.S. Golovin, Fuel elements of atomic reactors (Moscow: Gosatomizdat, 1962).

61. N.P. Agapova, A.A. Bochvar, A.S. Zaimovskiy, et al., Atomic Materials Technology, in: Soviet atomic science and engineering (Moscow: Atomizdat, 1967).

62. Yu.B. Khariton and Yu.N. Smirnov, Some myths and legends about Soviet atomic and hydrogen projects, in: [44].

63. V.S. Gubarev, M.F. Rebrov, and I.I. Mosin, The Bomb (Moscow: IzdAt, 1993).

64. B.G. Dubovskiy et al., Critical parameters of systems with fissile materials and

nuclear safety: A hand-book (Moscow: Atomizdat, 1966).

65. L.P. Sokhina, My memoirs of chemical production center "Mayak" (Chelyabinsk-40: Manuscript, 1993).

66. L.P. Sokhina, Ya.I. Kolotinskiy, and G.V. Khalturin, Price for plutonium: Documentary story about a chemico-metallurgical plant in its early years (Chelyabinsk-40: Manuscript, 1991).

67. L.R. Groves, Now We Can Talk About It (in Russian) (Moscow: Atomizdat, 1964).

68. V.I. Zhuchikhin, First atomic power station: Notes of a research engineer (Moscow: IzdAt, 1993).

69. Large Soviet Encyclopedia (BSE), 2nd ed., 1956, Vol. 51, Vol. 42.

70. Sketches of Sakharov's portrait as a scientist, Priroda, No. 8 (1992) (Special issue).

71. A.K. Kruglov and A.P. Rudik, Reactor production of radioactive nuclides (Moscow: Energoatomizdat, 1985).

72. Experimental Nuclear Physics, Ed. E. Segre (in Russian) (Moscow: Inostr. Lit., 1955, Vol. II).

73. G.A. Polukhin, First steps: History of production center "Mayak" (Chelyabinsk-40, 1993).

74. A.K. Kruglov and K. Tkach, Pioneers of secret atom, Atom-Pressa (Atomnaya energetika), No. 10 (1993).

75. V. Khokhryakov and S. Romanov, Radiation effects on lung cancer, NIIMB, No. 4 (1992).

76. V. Khokhryakov, K. Suslova, and A. Skryabin, Comparative analysis of plutonium accumulation in human organism, in: [75].

77. N. Koshurnikova et al., Radiation effects on the personnel of production center "Mayak", in: [75].

78. S.G. Kacheryants and N.N. Gorin, Pages of the history of "Arzamas-16" nuclear center (Arzamas-16: VNIIEF, 1993).

79. Biographies of memorable figures of Russia, X-XX centuries (Moscow: Moskovskiy rabochiy, 1992).

80. Yu.A. Romanov, The father of the Soviet hydrogen bomb, Priroda, No. 8 (1990).

81. Radiation safety standards NRB-76/87 and basic sanitary regulations for handling radioactive materials and other radiation sources OSP-72/87 (Moscow: Energoatomizdat, 1988).

82. H. Kissinger, Nuclear weapons and foreign policy (Transl. from English) (Moscow: Inostr. Lit., 1959).

83. R. Young, Brighter Than a Thousand Suns (Transl. from English) (Moscow, Atomizdat, 1960).

84. T. Cochran, W. Arkin, R. Norris, and G. Sands, Nuclear weapons of the USSR (Transl. from English) (Moscow: IzdAt, 1992).

85. V. Suvorov, The Lemon Country (Moscow: Sovet. Rossia, 1989).

86. Actinide Chemistry (Transl. from English) Eds.: G. Katz, G. Seaborg, and L. Morris (Moscow: Mir Publishers, 1991, vol. 1).

87. Z. Eklund. Oklo − Atomic reactor 1,800 million years ago, IAAE Bulletin, 1975, V. 17, No. 3 (in Russian).

88. Reactor Handbook, Physics (Transl. from English) (Moscow: Atomizdat, 1964).

89. V. Khapaev and E. Gudkov, Mystery of the Caucasus "pearl": German specialists – contribution to the domestic atomic industry creation, Atom-Pressa (Atomnaya Energetika), No. 12 (1993).

90. N.P. Galkin, et al., Production and concetration of uranium hexafluoride, in: Atomic science and engineering in the USSR (Moscow: Atomizdat, 1977).

91. C. Coen, Isotope separation, in: [52], Vol. 2 (1950).

92. A.M. Petrosyants, Life paths we were chosen by (Moscow: Energoatomizdat, 1993).

93. L.V. Matveev and E.M. Tsenter, Uranium-232 and its effect on the radiation situation in the nuclear fuel cycle (Moscow: Energoatomizdat, 1985).

94. E. Mikerin, Top-priority industry: Production of enriched uranium is still the most important sphere in the activity of Russia's Minatom, Atom-Pressa (Atomnaya Energetika), No. 4 (1993).

95. V.N. Prusakov, Advances in isotope separation, Jubilee session of the Scientific Council of Russia's Research Center "Kurchatov Institute" (Kikoin Conference) (Moscow: RNTs "Kurchatov Institute", 1993).

96. N.A. Koshcheev and V.A. Dergachev, Electromagnetic isotope separation and isotope analysis (Moscow: Energoatomizdat, 1989).

97. V.V. Gorin, G.A. Krasilov, A.I Kurkin, et al., Semipalatinsk test site: Chronicle of underground nuclear explosions and their primary radiation effects (1961–1989), Bulletin of Public Information Center for Atomic Energy, No. 9 (1993).

98. Research and development in reactor research centers (Moscow: RNTs "Kurchatov Institute", 1993).

99. V.V. Goncharov, Research reactors: Design and development (Moscow: Nauka, 1986).

100. Graves amd Packston, Critical masses of uranium-enriched systems (Transl. from English) (Moscow: Department of R&D Information and Exhibitions, 1958).

101. Atomic science and engineering in the USSR: Jubilee collection of papers (Moscow: Atomizdat, 1977).

102. Origin of the Soviet Atomic project: Role of intelligence, Voprosy Istorii Estesvoznaniya i Tekhniki, No. 3 (1992).

103. Sergo Beria, My father Lavrentiy Beria (Moscow: Sovremennik, 1994).

104. A.I. Yoirysh, I.D. Morokhov, and S.K. Ivanov, A-bomb (Moscow: Nauka, 1990).

105. Nuclear research reactors: Procedings of the US Commission on Atomic Energy (Moscow: Izd-vo Inostr. Liter., 1956).

106. A.K. Kruglov, History of the atomic science and undustry, Bulletin of the Public Information Center for Atomic Energy, No. 12 (1993).

107. A.D. Galanin, Introduction to the theory of thermal nuclear reactors (Moscow: Energoatomizdat, 1984).

108. B.M. Andreev, Ya.D. Zelvenskiy, and S.G. Katalnikov, Heavy hydrogen isotopes in nuclear engineering (Moscow: Energoatomizdat, 1987).

109. Physical Encyclopedic Dictionary (Moscow: Sovetskaya Entsiclopediya, 1960, V. 1, 1966, V. 5).

110. L.M. Yakimenko, I.D. Modylevskaya, and S.A. Tkachek, Water electrolysis (Moscow: Khimiya, 1970).

111. Kim Smirnov, In the memory of Academician Aleksandrov, Izvestiya, February 5 (1994).

112. A.I. Alikhanov, V.V. Vladimirskiy, S.Ya Nikitin, *et al.*, Experimental heavy-water reactor. Paper presented at the International Conference on the Peaceful Use of Atomic Energy, Geneva, 1955 (Moscow: Izd-vo AN SSSR, 1955).

113. V.V. Goncharov, Experimental basis of atomic reactor centers and its development, in: R&D in reactor research centers (Moscow: RNTs "Kurchatov Institute", 1993).

114. V.A. Kurnosov, 60 years of the All-Russia association "VNIIPIET", Atom-Pressa (Atomnaya Energetika), No. 1 (1994).

115. P.A. Petrov, Nuclear power units (Moscow-Leningrad: Gosenergo, 1958).

116. Chronicles of the motherland: Reactor for a submarine. Interview with E.P. Slavskiy, V.A. Chernavin, and A.P. Aleksandrov recorded by Captain S. Bystrov, Krasnaya Zvezda, October 21 (1989).

117. B.A. Pyatunin, *et al.*, The experience of nuclear reactor dismantling, Atomnaya Energiya, V. 69, No. 3 (1989).

118. V.I. Fetisov, Production center "Mayak" faces a new test – conversion, Atom-Pressa (Atomnaya Energetika), No. 31 (1993).

119. I.V. Gordeev, D.A. Kardashev, and A.V. Malyshev, Nuclear-physical constants, Reference book (Moscow: Gosatomizdat, 1963).

120. V.M. Gorbachev, Yu.S. Zamyatnin, and A.A. Lbov, Basic characteristics of isotopes of heavy elements (Moscow: Atomizdat, 1975).

121. V.E. Ivanov, I.I. Papirov, G.V. Tikhinsky, and V.M. Amonenko, Pure and extra-pure metals (produced by vacuum distillation) (Moscow: Metallurgiya, 1965).

122. S. Snegov, The creators: A historical novel about contemporaries (Moscow: Sovetskaya Rossia, 1979).

123. V.A. Kachalov, V.V. Pichugin, and A.V. Shchegelskiy, Chronological review of the Minsredmash history, Atom-Pressa (Atomnaya Energetika), No. 23 (1993).

124. L.R. Groves, Program for creating an atomic bomb; E. Teller, Program for creating a hydrogen bomb, in: [30].

125. A.P. Aleksandrov, Contribution of science to the benefit of the country: Papers and speeches (Moscow: Nauka, 1983).

126. E. Pyatov, Uranium reserves of Russia: Ways for developing the base of natural uranium mineral resources, Atom-Pressa (Atomnaya Energetika), No. 4 (1993).

127. 30 years of Czechoslovakian uranium industry (Praha, 1975).

128. TsNIIatominform express-information, No. 42 (1991).

129. O.L.Kedrovsky and D.I.Skorovarov, Mining and processing of uranium ore in the USSR, in: Atomnaya nauka i tekhnika SSSR (The USSR Atomic science and engineering) (Moscow: Energoatomizdat, 1987).

130. V.N. Mosinets, O.K. Avdeev, and V.M. Melnichenko, Wasteless technology for mining radioactive ore (Moscow: Energoatomizdat, 1987).

131. V.N. Mosinets and M.V. Gryaznov, Uranium mining industry and the environment (Moscow: Energoatomizdat, 1983).

132. G.A. Kovda, B.N. Laskorin, and B.V. Nevskiy, Uranium ore concentration technology, in: Soviet atomic science and engineering (Moscow: Atomizdat, 1967).

133. B.N. Laskorin, *et al.*, Integrated processing of uranium ore: Exploration and mining of radioactive raw materials, in: [90].

134. A.I. Kalabin, Mining of useful minerals by the underground leaching and other geotechnological methods (Moscow: Atomizdat, 1981).

135. O.S. Andreeva, V.I. Badiin, and A.N. Kornilova, Natural and concnetrated uranium: Radiation and sanitary aspects (Moscow: Atomizdat, 1979).

136. Appeal to mass media writers, editors, and journalists on the interpretation of radiation safety problem, Bulletin of Public Information Center for Atomic Energy, No. 8 (1993).

137. Power industry: Figures and facts (Moscow: TsNIIatominform, 1993).

138. TsNIIatominform express-information, Nos. 35–36 (1993).

139. B.V. Nikipelov, Reduced demand for natural uranium, Atomnaya energiya (Atomic Energy), V. 76, No. 3 (1994).

140. A.N. Kallistov, There was an emergency situation, Local Newspaper "Energiya", May 17 (1991), Elektrostal.

141. N.A. Volgin, Formation period: Reminiscences and contemplations (manuscript) (Moscow, 1993).

142. A.N. Kallistov, Two important events: Memoirs of the former director of machine-building plant, Local Newspaper "Energiya", June 7 (1991), Elektrostal.

143. History of Siberian atom – the history of peace: Figures, facts, reminiscences. Express SV, 1993.

144. G.N. Flerov, Research of the USSR Academy of Sciences on reactors with uranium-235, plutonium-239 and hydrogen moderator. Paper presented at the Session of the USSR Academy of Sciences Devoted to the Peaceful Use of Atomic Energy, July 1-5, 1955 (Moscow: Izd-vo AN SSSR, 1955).

145. G.Ya. Sergeev, V.V. Tutova, and K.A. Borisov, Uranium metal and some other reactor materials technology (Moscow: Atomizdat, 1960).

146. B.V. Brokhovich, V.I. Klimenkov, F.Ya. Ovchinnikov, et al., Dismantling of an experimental uranium-graphite reactor for isotope production after its four-year operation, in: Proceedings of the Second International Conference on the Peaceful Use of Atomic Energy, Geneva, 1958 (Moscow: Publishing House of the Chief Directorate on the Uses of Atomic Energy, USSR Council of Ministers, 1959, Vol. 2).

147. V.S. Gubarev, Chelyabinsk-40 (Moscow: IzdAt, 1993).

148. V.I. Ritus, If not me, then who? Priroda, No. 8 (1990).

149. A.G. Samoilov, A.I. Kashtanov, and V.S. Volkov, Dispersion of fuel elements: Design and serviceability (Moscow: Energoatomizdat, 1982).

150. Atom-Pressa (Atomnaya Energetika), No. 45 (1993).

151. L. Ievlev, Disintegration as a threat to safety, Posev, No. 2 (1992).

152. Yu.I. Koryakin, Russia's nuclear power deadlock: Is there a way out? Energeticheskoe Stroitelstvo, No. 7 (1993).

153. The state and problems of maintaining safety in Russia's nuclear facilities, Bulletin of Public Information Center for Atomic Energy, No. 4: 24 (1994).

154. Izvestiya, December 2 (1992).

155. Elimination of the effects of accidents caused by the disposal of radioactive waste, TsNIIatominform Review (Moscow: TsNIIatominform, 1992).

156. A.K Kruglov and Yu.V. Smirnov, Nuclear catastrophes: Consequences and prospects for the development of atomic industry (Moscow: TsNIIatominform, 1992).

157. A.A. Karatygin, Memoirs (1940–1978) (Obninsk: Manuscript, 1978).

158. Atom-Pressa (Atomnaya Energetika), June 20 (1992).

159. MacCulloch, Mills, and Teller, Safety of Nuclear Reactors, in: Proceedings of International Conference on the Peaceful Use of Atomic Energy, August 8–20, 1955. Vol. 13 (Moscow: Izd-vo Innostr. Liter., 1958).

160. A.K. Kruglov, History of the atomic science and industry, Bulletin of Public Information Center for Atomic Energy, No. 8: 56–57 (1993).

161. V.V. Goncharov, Research on the physics and engineering of research reactors: R&D in the reactor research centers (Moscow: RNTs "Kurchatov Institute", 1993).

162. Kippin and Whitman, Delayed neutrons, in: Proceedings of International Conference on the Peaceful Use of Atomic Energy, August 8–20, 1955 (Moscow: Izd-vo AN SSSR, 1958).

163. F. Kap, Physics and engineering of nuclear reactors (Transl. from German) (Moscow: Izd-vo Inostr. Liter., 1960).

164. Dietrich, Experiments on self-control phenomena and safety of reactors with water moderator, in: [159].

165. G. Medvedev, Nuclear sunburn (Moscow: Izd-vo Knizhn. Palata, 1990).

166. V.A. Sidorenko, Problems of safe operation of VVER reactors (Moscow: Atomizdat, 1977).

167. I.Ya. Emeliyanov, M.B. Egiazarov, V.I. Ryabov, et al., Physical start-up of the RBMK reactor, second block of the Lenigrad NPP, Atomnaya Energiya, V. 40, No. 2 (1976).

168. B.G. Dubovsky, A.K. Kruglov, et al., Nuclear safety in handling fissile materials, in: [101].

169. USSR State Committee for supervision of safe operation in the atomic industry (Adopted by the USSR Council of Ministers Decree No. 409 of May 4, 1984).

170. Yu.B. Khariton and Yu.N. Smirnov, Myths and reality of the Soviet atomic project (Arzamas-16, 1994).

171. Ya.P. Terletskiy, Mission "Interrogation of Niels Bohr", Problems of natural history and engineering, No. 2: 18–44 (1994).

172. V.N. Mikhailov, "I am a war hawk" (Moscow: Kron-press, 1993).

173. Yu.S. Zamyatnin, Paper presented on May 8, 1996, at Symposium on the History of the Soviet Atomic project, Dubna.

174. Pavel Sudoplatov, Secret service and the Kremlin: Notes of undesireable witness (Moscow: Geya, 1996).

175. Chernobyl: Catastrophe, Deed, Lessons and Conclusions. The 10th anniversary of the Chernobyl catastrophe (Moscow: Inter-Vesy Publishing House, 1996).

INDEX

Plant No. 813 *see* Center No.813
Plant No. 817 176–7
Plant V 69. 115. 187–8. 223, 225, 242–4
  additional product requirements 97–102;
  Block 9 facility 103–7; designing/mastering
  of technology 102–7; development of
  plutonium metallurgy 91–7; female
  personnel 103; interior exposure effects
  107–9
Platonov, P.A. 231
Plotkina, A.G. 145
plutonium 7–8, 15, 17–18, 22–3, 63–5, 215,
  218–19, 228, 258
  cleaning of 78–9; concentrated solution 69;
  extraction 74, 78; handling of 94–5;
  increased production of 124; metallic 69;
  metallurgy development 91–109; process
  technology 79; production of 71; prone to
  corrosion 94; secrecy surrounding 80–1;
  self-heating properties 94; studies of
  chemical properties 72; for weapons
  production 32–3
plutonium bomb 241
  and decision to copy American design
  116–17; explosion efficiency/effects 124–5;
  implosion principle 111–12; preparation/
  testing 119–24; and problems with neutron
  primer 117–19; security surrounding
  112–13; and structural materials impact
  compressibility studies on 113–16. *see also*
  atomic bomb; hydrogen bomb; nuclear
  weapons
Plutonium Center 46, 53, 82, 84, 103–4.
  176
Podolsk Heavy Machinery Works 169
Poido, M.S. 105–6
Polikovskiy, V.I. 139, 144
polonium 119
Pomeranchuk, I.Ya. 24, 43, 169, 226
Pomerantsev, G.B. 54, 62
Pontecorvo, B. 117
positron 12
Potsdam Conference (1945) 32
Pozdnyakov, B.S. 35, 47, 48, 53
Pozharskaya, M.E. 218
Pravdyuk, N.F. 47, 51, 167
Production Center No. 6 194
Production Center No. 9 192
production reactor A 230
  background 45–51; construction/start-up
  51–6; corrosion of aluminum pipes in 59;
  design of 47–8, 49–51; as First Research
  Base 62–5; problems in servicing 56–62; and
  radiation exposure 65–8; radiation safety
  during rehabilitation 251–2. *see also*
  F-1 reactor; nuclear reactor
protons 2, 7
Purtova, T.A. 233

Rabotnov, S.N. 54, 81
radiation,
  accidents 59–60, 89, 242–4, 249–50; average
  doses received by personnel 61; control
  of 62; efforts to reduce exposure to 81–5;
  environmental aspects 205–11; exposure
  to 65–8, 79–80, 87–8, 107–9; monitoring
  of 88–9; mortality rates 61, 66–7, 84, 87,
  108–9, 242; "pancake" accidents 58; safety
  aspects 36, 43, 53–4, 75, 235–56
Radiation Laboratory 62
Radiation Safety Standards (RNB) 253
radioactivity 11, 200
  annus mirabilis 12; early period research 1–4;
  tables of characteristics 4
Radiochemical Plant B 97, 102, 242, 244
  background 69–75; design/construction 75–8;
  efforts to decrease personnel exposure 81–5;
  hazardous effects of Radwaste disposal at
  natural water basins 85–9; initial operation
  period 78–81
radiochemistry 13, 257
radioelements 1
Radiological Laboratory 70
radiometry 72
radium 4, 70
  extraction of 14; production of 5; search for 5;
  and treatment of cancer 14
Radium Commission 4
Radium Institute of the Academy of Sciences
  (RIAN) 6, 11–14, 25, 63, 69, 71–2, 74, 81,
  118, 122, 216, 218, 253, 257
Radium Institute (Austria) 217
Radium Institute (Paris) 1
Radkevich, I.A. 181
radon 4
radwaste 74
  disposal of 85–9
Raikham, M.L. 145, 147
Ratner, A.P. 71–3, 75, 78, 84, 243
RBMK reactor 233
RDS-1 ("Russia did it on her own") 123
Reichmann, R. 131, 136
Research Institute of Atomic Reactors 175
Research Institute for Chemical Machine
  Building (NIIKHIMMASH) 48, 49, 224
Research Institute of General and Inorganic
  Chemistry (IONKh) 102, 105–6
Research Institute of Geochemistry and
  Analytical Chemistry (GEOKhI) 74, 105,
  122, 216
Research Institute-42 141
Reshetnikov, F.G. 105, 222
RIAN *see* Radium Institute of the Academy of
  Sciences
Riel, N. 130–1, 217, 260
Ritus, V.I. 101, 228
Rodionov, M.P. 145, 147